THE MAKING OF A TROPICAL DISEASE

JOHNS HOPKINS BIOGRAPHIES OF DISEASE
Charles E. Rosenberg, Series Editor

Randall M. Packard, *The Making of a Tropical Disease:
A Short History of Malaria*

Steven J. Peitzman, *Dropsy, Dialysis, Transplant:
A Short History of Failing Kidneys*

THE MAKING
of a TROPICAL
DISEASE

❖ ❖ ❖

A Short History of Malaria

Randall M. Packard

THE JOHNS HOPKINS UNIVERSITY PRESS
Baltimore

Johns Hopkins Paperback edition, 2011
2 4 6 8 9 7 5 3 1

The Johns Hopkins University Press
2715 North Charles Street
Baltimore, Maryland 21218-4363
www.press.jhu.edu

*The Library of Congress has catalogued the hardcover edition of
this book as follows:*

Packard, Randall M., 1945–
The making of a tropical disease : a short history of malaria /
Randall M. Packard.
p. ; cm. — (Johns Hopkins biographies of disease)
Includes bibliographical references and index.
ISBN-13: 978-0-8018-8712-3 (hardcover : alk. paper)
ISBN-10: 0-8018-8712-7 (hardcover : alk. paper)
1. Malaria—History. I. Title. II. Series.
[DNLM: 1. Malaria—history. WC 750 P119m 2007]
RC160.P33 2007
616.9'362009—dc22   2007011828

A catalog record for this book is available from the British Library.

ISBN-13: 978-1-4214-0396-0
ISBN-10: 1-4214-0396-X

CONTENTS

Disease is a fundamental aspect of the human condition. Ancient bones tell us that pathological processes are older than humankind's written records, and sickness still confounds our twenty-first century's technological pride. We have not banished pain, disability, or the fear of death even if we die, on average, at older ages, of chronic and not acute ills, in hospital or hospice beds, and not in our own homes. Disease is something men and women feel. It is something in our bodies—but also in our minds. Disease demands explanation; we think about it and we think with it. Why have I become ill? And why now? How is my body different in sickness from its quiet functioning in health? Why in times of epidemic has a whole community been scourged?

Answers to such timeless questions necessarily mirror and incorporate available ideas and assumptions. In this sense, disease has always been a social and linguistic as well as biological entity. In the Hippocratic era, physicians—and we have always had them with us—were limited to the evidence of their senses in diagnosing a fever, an abnormal discharge, or seizures. Classical notions of the somatic basis for such alarming symptoms necessarily reflected and expressed contemporary philosophical and physiological notions, a speculative world of disordered humors and "breath." Today we can call for understanding upon a variety of scientific insights and an armory of diagnostic and therapeutic practices—tools that allow us to diagnose ailments unfelt by patients and imperceptible to the doctor's senses. In the past century disease has become increasingly a bureaucratic phenomenon, as well, as sickness has been defined and in a sense constituted by formal disease classifications, treatment protocols, and laboratory thresholds.

Sickness is also linked to climatic and geographic factors. How and where we live and how we distribute our resources all contribute to time- and place-specific incidence of disease. For example, ailments such as typhus fever, plague, malaria, dengue, and yellow fever reflect specific environments that we have shared with our insect contemporaries. But humankind's physical circumstances are determined in part by culture—especially agricultural practice. Environment, demography, ideas, and applied medical knowledge all interact to create particular ecologies of disease at particular moments in time.

Disease is thus historically as much as biologically specific. Or perhaps I should say that every disease has a unique past. Once discerned and named, every disease claims its own history. At one level biology creates that idiosyncratic identity. Symptoms and epidemiology as well as cultural values and scientific understanding shape responses to illness. Some writers may have romanticized tuberculosis, but as the distinguished medical historian Owsei Temkin noted dryly, no one had ever thought to romanticize dysentery. Tuberculosis was pervasive in nineteenth-century Europe and North America and killed far more people than cholera did, but it never mobilized the same widespread and policy-shifting anxiety. Unlike tuberculosis, cholera killed quickly and dramatically and was never accepted as a condition of life. Sporadic cases of influenza are normally invisible, indistinguishable among a variety of respiratory infections; waves of epidemic flu are all too visible. Syphilis and other sexually transmitted diseases, to cite another example, have had a peculiar and morally inflected attitudinal history. Some diseases such as smallpox or malaria have a long history, others like AIDS a rather short one. Some have flourished under modern conditions; others seem to reflect the realities of an earlier and less-economically-developed world.

These arguments constitute the motivational logic behind the Johns Hopkins Biographies of Disease. Each historically visible entity—each disease—has a distinct history. Biography implies chronology and narrative—a movement in and through time. Once inscribed by name in our collective understanding of medicine, each disease entity becomes a part of that collective understanding and, thus, inevitably shapes the way in which individuals think about their own felt symptoms and prospects for future health.

No disease illustrates the complex interdependencies that shape disease incidence and experience better than malaria; like infant mortality it is an index to a society's material conditions and social arrangements. Malaria is multifactorial, exquisitely sensitive to particular environmental circumstances and social and economic relationships. Nevertheless, since classical antiquity, physicians and observers have associated it with environmental circumstances, with marshes, with dampness, and with low-lying lands. But geography is not malaria's one-dimensional destiny; in the past it has flourished in both old England and New England. Poverty and inequality as much as geography and climate make malaria a "tropical" disease. Randall Packard has demonstrated in this wide-ranging synthesis how much malaria is both actor and acted upon in agricultural history, in the distribution of political and economic power, in imperial relationships, and in the movement of populations. Although a significant factor, it has never been an independent variable, even if historians have associated it in retrospect with the declining fortunes of classical Greece and Rome. Our understanding of the role of the anopheles mosquito as vector or the availability of quinine and its descendants as tools for malaria's clinical management has constituted only one factor in the disease's changing fortunes. Accumulating medical knowledge has created choices, not dictated them. The multifaceted history of malaria illuminates fundamental aspects of human history.

Charles E. Rosenberg

Mulanda

I have spent much of the past 20 years researching and writing about the history of malaria, a disease caused by a parasite that is transmitted through the bite of a female *Anopheles* mosquito. My interest in the disease began in the late 1960s when I first experienced it directly. I was part of a team of Peace Corp Volunteers sent to southeast Uganda to eradicate an infectious eye disease known as trachoma. We were stationed at a small rural clinic in Mulanda, a village some 13 miles by dirt road from the town of Tororo. The clinic served a population of several thousand men, women, and children, living in dispersed homesteads, scattered over some 125 square miles of boulder-studded hills separated by rivers and swamps.

Because malaria was endemic in the area, we took weekly doses of chloroquine as a prophylactic. The drug attacks the parasite circulating in the bloodstream and suppresses the infection. This simple precaution protected my fellow volunteers for the two years they were in Uganda. However, six months into my stay I managed to miss a dose and, as luck would have it, I had been bitten by an infected mosquito. Without the suppressive effect of the chloroquine, the parasites were able to proliferate in my bloodstream. I developed the classic symptoms of the disease, with intermittent bouts of high fever and profuse sweating, followed by chills that wracked my body and persisted no matter how many blankets I piled on. But what I remember most about my first encounter with malaria was the extreme pain in my head, worse than any headache I had experienced before or since. After a few days I was taken to the district hospital in Tororo and treated with injectable chloroquine. The fever and chills subsided, and while I remained weak and exhausted for days, I eventually recovered. Malaria is a curable disease if treated promptly and properly. I was fortunate enough to receive such treatment. Many of those living around me in Mulanda were not.

A good number of the patients who attended the clinic at which I worked suffered from malaria, especially toward the end of the long rains, which ran from March to May. Today malaria accounts for nearly 40 percent of outpatient visits in Uganda. Most of the malaria cases then, as now, involved young children. They often arrived in very poor condition, cradled in their mothers' arms, hardly moving except for the rapid breathing caused by their high fevers. I would often see them waiting to be examined, sitting in the shade of a large frangipani tree that grew by the side of the clinic. Sometimes they would come at night, and I would hear the clinic watchman calling for the medical attendant, who lived in a house next to mine in the clinic compound. If they were very ill, they would be given injectable chloroquine; otherwise, a few chloroquine tablets and aspirin. Chloroquine injections were not always available, however, and even the chloroquine tablets could run out, especially during seasons of intense transmission. When this happened, the mothers were simply given aspirin and told to return the next day, when there might be more medicine. They could also buy chloroquine tablets at a local shop, if they could afford them. Sadly, not all of these children survived. Those who did not live often suffered from cerebral complications caused by infection with *Plasmodium falciparum* malaria, which, if left untreated, reproduces in huge numbers, destroying red blood cells and clogging capillaries, including those leading to parts of the brain. More than a few of these children died at the clinic. For two years, the wailing cries of grieving mothers punctuated my days and nights in Mulanda.

Why did a curable and largely preventable disease kill so many children? I directed this question to the senior medical attendant at Mulanda as we sat together one evening under the frangipani tree, sipping warm beer. He said the parents waited too long to bring their children to the clinic. I asked why this was so. Wasn't medical treatment provided free of charge? He replied that the parents often took their children to a traditional healer before coming to the clinic. I later learned that this was true, in some cases. There was an indigenous category of disease that corresponded to the high fever and headaches associated with cerebral malaria. Yet the delay could also occur because the parents were away when the child became ill. Sometimes parents would try to treat their sick child using medications purchased at a local shop. Lack of knowledge about the proper dosage, or the poor quality of the drugs, or the parent's attempt to make the drugs last longer by using less than the recommended dosage made home treatment a risky enterprise. Finally, parents who lived miles from the clinic hesitated to spend money on the bus fare in order to bring the child for treatment until he or she was very sick, especially because the needed medications were not always available at the clinic. For whatever reason, these children did not receive prompt and appropriate treatment.

Trachoma was also a preventable and treatable disease. The simple observation of sanitary rules, washing hands and not sharing face cloths, could greatly reduce transmission. The application of tetracycline three times a day to the infected eye and a 12-week regimen of long-acting sulfa drugs cured the disease. Yet many people contracted the disease, and many of those we treated returned reinfected. I once treated a young woman in the early stages of trachoma in her left eye. She returned for her sulfa tablets each Monday for three weeks as instructed. The combination of drugs and tetracycline did their work, and her eye improved quickly. She then disappeared for a month only to return with both eyes infected. I asked her why she had stopped coming for treatment. She replied that her mother had died, and she had had to go to her funeral. Her family lived in Busoga, an adjoining district with a high prevalence of trachoma. It was not uncommon for people to spend weeks at a funeral, nor was it uncommon for them to disappear from treatment during this time.

There was no easy solution to this problem. We were not allowed to give patients multiple doses of sulfa to take with them because there was a black market for the drugs, which could also be used to treat other ailments, including chlamydia and gonorrhea. There was no trachoma eradication program in Busoga and thus no way for a patient to continue treatment there. People could not afford to return each week for their treatment from so far a distance. There was also no way to avoid the death of a parent or the social obligations that followed such losses. The woman who was lost to treatment was not stupid or unaware of what she needed to do to regain her health. But her ability to do so was compromised by the unpredictability of life events; the cultural norms of her community; the inadequacies of the Ugandan health system; the absurdity of creating a disease eradication program in a region of the country surrounded by other heavily infected areas where no control program existed; and her lack of financial resources to continue treatment at a distance. When it came to completing treatment, the odds were stacked against her.

As I traveled from homestead to homestead, following up with the patients we had seen and treated at the clinic, and providing public health information, I came to realize that the failure of public health efforts to prevent and treat both malaria and trachoma resulted from more fundamental causes. The families I visited had few resources. Clean water to wash with was not readily available. Women and children often had to travel considerable distances to find a protected borehole from which clean water could be drawn. It was often easier to collect a bucket of water from a local stream or from the swamps that separated villages, even though the water was often badly polluted. Purchasing bedclothes and face cloths for each member of the family to prevent the spread of trachoma, let alone screens for windows to keep out malaria-infected mosquitoes, was beyond the means of many families. When people became ill, they often lacked the resources to obtain medical care in a timely manner. In the case of trachoma, traveling to the clinic once a week for twelve weeks was a heavy burden on those who lived miles away.

These relatively small expenditures were beyond the means of many families because most earned less than the equivalent of one dollar (U.S.) per day. Their income came largely from cotton grown on small plots of land. It is said that cotton is the mother of poverty in Africa, and this was certainly true in this corner of Uganda. The price of cotton was and is determined by international commodity agreements. Because cotton is grown over many parts of the globe, the supply generally exceeds demand, keeping prices low. On top of this, the government fixed the price paid each season to Uganda farmers well below international cotton prices so that it could extract income to be used for development. Unfortunately, most of these extracted funds were devoted to building roads, hospitals, and convention centers in urban areas. Little of the money made its way back to the rural villages from which it came. Despite this inequitable arrangement, local farmers needed money to pay taxes, and if possible to pay for their children's school fees, and had few alternative ways to earn cash.

I once asked an agricultural officer how the farmers in Mulanda could increase their production and thus their income. He noted that few farmers in the area employed intensive farming methods, preferring to plant a field for one season and then move on to another parcel of land the following season, leaving the first field to lie fallow for two or three seasons so that the soil would regain its fertility. If they applied fertilizers, he pointed out, they could reduce fallow periods, plant more cotton, and increase their income. Unfortunately, few farmers could afford fertilizer. This was not simply because they lacked money. Cotton fertilizer was sold in large bags, which contained enough fertilizer to optimally cover one acre of land. Few farmers possessed an acre of land. Most had less than half an acre. In order to purchase fertilizer they would have to pay for more than they could use. The fertilizer did not keep well between seasons, and so much of it could be lost. The only way for a farmer to employ fertilizer efficiently was to cooperate with a neighbor.

This discussion led me to write a proposal to the head of the local U.S. Agency for International Development mission in Kampala, requesting funds to begin an agricultural cooperative that would allow farmers to pool resources in order to purchase fertilizers and other inputs to increase production. My request was turned down, and I was politely told to return to treating patients. The idea that the solution to the area's malaria and trachoma problems might lie in the transformation of agricultural production methods was not appreciated by USAID. It was nonetheless true. Both malaria and trachoma took a toll on the people of Mulanda because they could not afford to purchase the simple commodities and medicines they needed to prevent and cure these diseases. In the end, I obtained some funds from USAID to protect springs, increasing the availability of clean water. But this did not increase my neighbors' income, and the lines of patients with trachoma, malaria, and other health problems continued to form outside the Mulanda clinic.

More than thirty-five years later I remain haunted by malaria, dying children, and grieving parents. This preventable disease continues to take the lives of children at a horrific rate, despite the billions of dollars that have been spent on finding ways to control it. Worldwide, between 350 and 500 million people are infected with the disease every year, and between 2 and 3 million die from it. Of all the deaths, 90 percent occur in Africa, where it is estimated that one child dies from the disease every 30 seconds. Malaria is also a major contributor to overall childhood mortality; in fact, the presence of malaria parasites nearly doubles overall mortality.[1] Pregnant women with malaria are subject to miscarriages and often give birth to low-weight babies, who suffer from high rates of infant mortality. People with malaria are more susceptible to other infections, most notably today HIV/AIDS. Finally, where malaria does not kill, it disables. Unlike yellow fever, measles, or polio, malaria infection does not produce a natural immunity to the disease. Individuals can be infected over and over. While recurrent infections can, over time, produce resistance to the disease, limiting the severity of subsequent infections, this resistance does not prevent reinfection. Recurrent infections can lead to chronic anemia and enlargement of the liver and spleen, leaving infected populations weakened and unable to perform normal daily tasks. This can have significant economic consequences for individuals, communities, and nations.

Over the years since I left Mulanda, I have developed a deeper interest in African history and the history of disease. Although I have researched and written on a range of topics, I have repeatedly been drawn back to the problem of malaria. My work, moreover, continues to be guided by the insights gained in Mulanda as I continue to explore the links that exist between the occurrence of this disease and the social and economic environment in which it occurs.

This book is an effort to record what I have learned from these inquiries: to take a look at the long history of malaria, from its

earliest occurrence in Africa and Southeast Asia to the present; and
to use this history to explain why malaria continues to plague millions of people across the globe, and why our efforts to eliminate
the disease have been largely unsuccessful. A great deal has been
written about malaria, almost entirely focused on specific regions
of the globe. This book is the first attempt to link these histories
together into a global narrative.

In the end the book makes a simple point. Malaria policy
needs to be informed by history. The history of malaria tells us
that malaria cannot be understood or eliminated independently
of changes in the societal forces that drive it. This is not to argue
that current and past malaria control efforts have had no effect
on reducing the burden of malaria. Nor do I believe that malaria
cannot be controlled before social and economic impediments to
health are completely removed. Rather I argue that the array of
biomedical weapons mobilized in the war against malaria needs to
be joined with efforts to understand and improve the social and
economic conditions that drive the epidemiology of the disease.
Only by making this connection, by learning the lessons of history, will we be able finally to stop the dying and to silence the
wailing.

THE MAKING OF A TROPICAL DISEASE

Constructing
a Global Narrative

❖ ❖ ❖

ARCHANGEL

Archangel is a port city in northern Russia located on the River Dwina, six miles from the White Sea. The city, established in the sixteenth century, sits some 125 miles south of the Arctic Circle, on roughly the same latitude as Fairbanks, Alaska. Daily temperatures in Archangel average around 10 degrees (Fahrenheit) in December and January and only in the 50s during the height of summer. For six months of the year, access to the city by water is only possible with the help of icebreakers. Russia → malaria?

Archangel would seem to be an unlikely place to begin an investigation into the history of malaria, a classic tropical disease. Yet, in 1922–23 an epidemic of malaria struck the citizens of this arctic port, causing an estimated 30,000 cases and more than 1,000 deaths. The epidemic that hit Archangel was part of a much larger regional epidemic that ravaged areas of Central Asia, the Caucasus, and the Volga basin, claiming an estimated 600,000 lives.[1] The history of this epidemic is a useful starting point for our inquiry, for it reveals a good deal about the wider history of this disease and its persistence.

Before the epidemic of 1922–23, malaria had been a serious health problem in Russia but one restricted mainly to the southern subtropical areas of the country. In 1922–23 a series of adverse climatic events, in combination with a succession of social, political, and economic crises, caused malaria to expand northward to the Arctic Circle. Malaria specialist Lewis Hackett described the conditions that brought on this epidemic in his 1934 classic, *Malaria in Europe:*

> In the middle Volga basin there had been an almost complete lack of rain for two successive years. The crops suffered the first year and in the second year they were destroyed. All the domestic animals died, either from lack of food or because they were sacrificed to the hunger of the population. Great masses of the people emigrated to more fortunate regions, where they became infected with new kinds of malaria, and in the mean time the immunity of those who stayed at home fell to a low level. The following year a great flood of the Volga inundated kilometers of plain along its left bank, and the receding waters in the summer turned all the depressions in the steppes into marshes which persisted throughout the breeding-season. On this physically reduced population, destitute of any biological defense either of domestic animals or of acquired immunity, descended the hordes of anophelines, and to add to the tragedy returning emigrants, who had heard that the land was again productive, brought their new parasites.[2]

Hackett's description highlighted the proximate climatic and social developments that led to the 1922–23 epidemic: the drought that damaged crops and undermined peoples' resistance to disease,[3] the subsequent flooding that produced breeding conditions for local anopheline mosquitoes, and the mass migration of people in search of food who returned infected with malaria. However, we need to look as well at the political and economic context in which these developments occurred in order to understand why they had such a dramatic impact on the health of the people of this region.

The 1922–23 malaria epidemic was, in fact, the culmination of a series of social and economic upheavals generated by the Russian Revolution and Civil War between 1917 and 1920. These events devastated agricultural production in Russia before the onset of drought. Only two-thirds of the land under cultivation at the beginning of World War I was being farmed at the Civil War's end.

Without fertilizers, machinery, draft animals, and even sufficient tools, Russian farmers in 1920 reaped barely two-fifths of what they had harvested in 1916. To make matters worse, the meager food reserves of many farmers and their families were seized by Bolshevik grain-requisitioning detachments during the Civil War. The food crisis was heightened by the demobilization of the Red Army's 5 million men, many of whom flooded back to the countryside, adding to the population that needed to be fed. The disruption of the Russian economy also led to the closing of a number of factories, including one that made bags for shipping grain. This closure limited the government's ability to provide famine relief to the hardest hit areas.[4] Finally, a Western blockade on Soviet shipping, following the Revolution, restricted the availability of quinine for treating malaria.[5] *environment + health*

Thus, the drought that occurred in 1921 exacerbated an already dire situation resulting from the political turmoil of the preceding four years. Had the Civil War not severely damaged the country's economic infrastructure and the peasants been allowed to retain the small food reserves they had accumulated before the drought, the famine might have been avoided or at least limited in its effects. Similarly, had large numbers of peasant families not been forced to seek out alternative food sources in regions where malaria was endemic, they would not have been exposed to new forms of malaria, which they introduced into the Volga region upon their return.[6] Malaria expanded northward through the Volga region and into the temperate regions of northern Russia in 1922–23 because it fed off the conditions of intense social, political, and environmental disruption that marked the history of this area between 1916 and 1921.

BENGAL

Three thousand miles to the south and east of Archangel lies the Ganges River delta, which dominates the physical landscape of West Bengal and Bangladesh. Here, we are in a tropical environment. Much of the area is inundated with floodwaters during the monsoon period from June to October. With its heavily irrigated lands, the region would appear to provide an ideal breeding ground for malaria. And in fact during the second half of the

nineteenth century, the western half of this region experienced
high levels of malaria morbidity and mortality. During the 1870s
it was estimated that malaria struck 75 percent of the inhabitants
of some villages in western Bengal, killing as much as 25 percent
of their population.[7]

Yet for much of the eighteenth and early nineteenth centuries not many people died of malaria in this region. The same
areas that were hard hit by malaria at the end of the century were
known as healthy and prosperous at the beginning of the century. In 1760 officers of the East India Company described Burd-
wan, located in West Bengal, as the most productive district in
the province of Bengal. Malaria was controlled, and the general
prosperity of West Bengal was ensured by the seasonal flooding of
the Ganges tributaries. Although flooding is often associated with
increases in malaria, this annual inundation suppressed the breeding of local species of anopheline mosquitoes. The annual floods
also brought enriching soils that fertilized the land and ensured
successful crops.[8]

During the second half of the nineteenth century, however,
the British transformed the physical landscape of much of western Bengal. The process began with the construction of embankments, built for roads and railways but which also served to protect the region from the danger of excessive flooding in the event
of heavy rains. Once this transportation network was in place,
aspiring Indian capitalist landowners sought to take advantage of
it by expanding the production of rice for market. To achieve this
end, they dammed rivers and streams to provide irrigation for
their lands. Unfortunately, the various embankments and dams
transformed the region's ecology in ways that both undermined
agricultural production and promoted the spread of malaria. Rivers silted up, and the areas that had been flushed out by annual
inundations became waterlogged, with swamplike conditions providing ideal breeding grounds for the introduction of new species of anopheline mosquitoes.[9] *Anopheles philippinensis,* which
had been restricted to relatively narrow zones in Bengal as long as
free flooding prevailed, proliferated in large numbers in the transformed swampy environment. The failures of annual floods to
deposit rich silts also reduced the fertility of the soil, leading to a

[handwritten: → scenario made the society even more susceptible to malaria]

decline in agricultural productivity. The combined decline in agricultural production and increase in malaria epidemics led farmers to abandon the land, leading to its further ecological deterioration and further rises in malaria.

PALM BEACH

Nine thousand miles to the west of Bengal in Palm Beach County, Florida, a 46-year-old man reported to the emergency department of a local hospital on July 22, 2003.[10] He had a three-day history of fever, headache, chills, anorexia, nausea, vomiting, dehydration, and malaise. He was treated with intravenous fluids and discharged with levofloxacin, a drug used to treat pneumonia and chronic bronchitis, as well as infections of the urinary tract and kidneys. On July 24 he returned to the emergency department with worsening symptoms and was admitted with a diagnosis of pneumonia. On July 25 *Plasmodium vivax* malaria was identified on a blood smear. The patient, a construction worker who reported working outside, recovered after treatment with doxycycline, quinine, and primaquine. *[handwritten: misdiagnosis!]*

Three weeks later, on August 15, a 32-year-old man was admitted to the same hospital with a 33-day history of fever, chills, headache, vomiting, and intermittent sweating. He had consulted several physicians for his symptoms and had been treated unsuccessfully with azithromycin and prednisone. On the day of admission, *P. vivax* was identified on a blood smear. The patient recovered after treatment with doxycycline, quinine, and primaquine. He reported having played golf and tennis in the evenings. *[handwritten: P. vivax = malaria parasite]*

Four days later, on August 19, a 45-year-old man visited the emergency department with a two-day history of fever, chills, anorexia, pains in the joints, and diarrhea and was discharged on a regimen of ibuprofen. The patient returned to the emergency department again on August 21 with the same symptoms, was evaluated, and discharged. On August 22, he again returned with worsening symptoms and mental confusion and was admitted; a blood smear demonstrated the presence of *P. vivax*. He recovered after treatment with chloroquine and primaquine. Before being infected, the patient had slept in a homeless camp in a wooded area near a canal; he reported using an insect repellent while there.

[handwritten: → what travelers take to avoid malaria]

He was one of an estimated 4,000 homeless people living in the county.

A construction worker, a golfer, and a homeless person were three of eight persons who appeared to have been infected locally with malaria. The index case from which these eight cases derived was not identified, although the Centers for Disease Control (CDC) concluded that a migrant worker or international traveler might have been involved. Many migrants working in the Palm Beach area come from areas of Latin America in which malaria remains a health problem. If these migrants had been repeatedly infected with malaria and had developed some resistance to it, they may have remained asymptomatic or experienced milder symptoms than would people with no previous exposure to the disease. Such migrants would have been less likely to seek medical help. They may also have had less access to care. Either way, their infections could have gone undetected, providing extended opportunities for them to infect local mosquitoes. In addition, the hospital staff, inexperienced in working with the disease, failed to correctly identify the infections of the eight cases. The delay in making the diagnosis allowed these sick patients to continue to infect local mosquitoes.

Palm Beach County was riddled with drainage ditches and canals, which were prime habitats for *Anopheles quadrimaculatus* mosquitoes, which are capable of transmitting malaria. This type of mosquito normally did not occur in large enough numbers to be a nuisance, and thus did not generate demand for mosquito abatement services. Moreover, most ditches and canals were kept relatively free of aquatic weeds, which sheltered the mosquito larvae in the water, and kept them safe from predation by fish. However, when neglected, these ditches became clogged, greatly increasing the breeding potential of *A. quadrimaculatus* mosquitoes. This increase, in turn, facilitated malaria transmission in 2003.

The CDC investigated the outbreak and concluded that it demonstrated the potential for reintroduction of malaria into the United States despite intensive surveillance, vector-control activities, and local public health efforts to educate clinicians and the community.

LESSONS

[handwritten: interesting case comparing all 3]

The disparate histories of malaria in Archangel, Bengal, and Palm Beach share three characteristics that have been central to the history of malaria. First, they all involved an environmental change *[handwritten: ✶]* that made possible the growth of anopheline mosquito populations. Specifically, in each case, changes in the distribution and flow of water across the land sparked an upsurge in anopheline production and malaria transmission. Water is essential for the transmission of malaria. Female anopheline mosquitoes lay their eggs in water, and these eggs advance through several stages of development within water before emerging as adult mosquitoes. Not any body of water will do. Anopheline mosquitoes are finicky. Each species prefers specific aquatic conditions. Some favor standing pools of water, others the edges of flowing streams. Whether the water is exposed to sunlight is important, as is the nature of vegetation in the water and the degree of salinity and pH level. The ability of a species of *Anopheles* to breed, reproduce, and survive depends on the availability of the correct conditions, though some species, such as the *Anopheles quadrimaculatus*, which is found over the eastern half of the United States, can adapt to a range of conditions.

Changes in the distribution and flow of water across the land may produce conditions that support the production of specific species of *Anopheles*, while discouraging the production of other species. These changes can have important consequences for the epidemiology of malaria. In Russia, extensive flooding in 1922 resulted in the creation of standing pools of water over wide areas of the upper Volga River basin, which resulted in an explosion of the population of *A. maculipennis* mosquitoes. In Bengal, the building of embankments restricted the flow of floodwaters, creating waterlogged environments that permitted *A. philippinensis*, a highly efficient malaria transmitter, to replace less efficient vectors. In Palm Beach, a breakdown in routine sanitation resulted in vegetation being allowed to grow in drainage canals, which produced an environment hospitable to the breeding of *A. quadrimaculatus* mosquitoes.

In addition to the presence of malaria-transmitting mosquitoes,

[handwritten: ▷ environmental factors play huge role in malarial transmission]

the epidemics in Archangel, Bengal, and Palm Beach depended on the existence of human populations who were susceptible to infection and lived or worked in conditions that exposed them to the bites of these mosquitoes. Human susceptibility to malaria is variable. As we will see in chapter 1, individuals can acquire resistance to specific species of malaria through repeated exposure. This resistance does not prevent infection but limits the severity of the infection. Whole populations may develop genetic immunities through long-term exposure. Both forms of resistance temper the impact of malaria. At the other extreme, in populations with no or limited prior exposure, the disease produces higher parasite levels and more severe symptoms. Immunologically naive populations are also more likely to infect mosquitoes. The susceptibility of individuals can also be affected by an individual's overall health and nutrition. Malaria attacks red blood cells and can cause severe anemia. When combined with general malnutrition or other infections, this anemia can be fatal. *cause of death 9,*

The people living in Archangel, in all likelihood, had never been exposed to malaria and thus were particularly susceptible. Equally important, the famine that preceded the epidemic caused widespread malnutrition that contributed to the lethality of the disease. In Bengal, malaria was endemic, and the population may have had some resistance to the disease. Yet the changes in water flow, which encouraged the breeding of *A. philippinensis* mosquitoes, also undermined the productivity of the land, leading to hunger. The effects of malnutrition may account for the high levels of mortality experienced by local villagers. Also, the fact that the villagers lived in close proximity to their fields and thus to the breeding grounds increased their exposure to the *A. philippinensis* mosquitoes.

The residents of Palm Beach, like those of Archangel, had no experience with the disease and became very sick. We know little about their overall health status, though the homeless person may have been poorly nourished. Like the residents of Archangel and Bengal, the eight affected persons lived or worked in conditions that exposed them to the bite of an infected mosquito but for a variety of reasons. For the construction worker, it was the conditions of his employment; for the homeless man, his lack of housing; and for the golfer, his practice rounds late in the day, at an hour when anopheline mosquitoes prefer to feed. Their circumstances were different, but they all engaged in activities, not always of their own choosing, that exposed them to malaria.

The third characteristic that the epidemics in Archangel, Bengal, and Palm Beach shared, in addition to the changing flows of water and exposed susceptible human populations, was the presence of malaria parasites. In Bengal these parasites were endemic to the region. In Archangel the parasites were foreign intruders, some of which may have been introduced into the region by returning refugees who had fled the famine and were infected in the south. Finally, the residents of Palm Beach were exposed to parasites that were in all likelihood introduced by a migrant worker from Latin America or an international traveler. Malaria parasites

The epidemics in Archangel, Bengal, and Palm Beach provide us with two more important lessons that are critical for understanding the history and persistence of malaria. First, malaria is not a disease that is limited to the tropics. The combination of conditions that gave rise to these malaria epidemics—changing flows of water, the existence of an exposed population, and the presence of malaria parasites—can and have occurred over wide areas of the globe. Although malaria is concentrated currently in tropical and subtropical regions, this distribution is a relatively recent development, dating from the middle of the twentieth century. The distribution of malaria in the middle of the nineteenth century included temperate regions of North America stretching northward into Canada; western and northern Europe; and Central and East Asia. Well over 50 percent of the world's population was at risk of malaria, and of those who were affected by the disease, 1 in 10 could expect to die from it.[11] x not limited to a certain area/u

Moreover, the opportunities for malaria to reemerge as a health problem in the North, despite the presence of well-developed public health and medical services, are real, even though the likelihood of an epidemic on the scale of Archangel or Bengal occurring in North America or western Europe is remote. Levels of sanitation, housing, and public health surveillance in most of the countries in these regions discourage the occurrence of anything more than localized clusters of cases. Still the convergence of inter-

many factors = influential

national travel, transnational labor migration, and homelessness, in the presence of mosquitoes capable of transmitting malaria and the breakdown of sanitation, raises the possibility of future recurrent malaria outbreaks. This possibility may also be heightened by the impact of global warming on the distribution and reproductive capacity of both malaria vectors and parasites, though as yet there is no clear evidence that warmer global temperatures are producing increases in the disease. Finally, it is worth noting that had the people in Palm Beach been infected with the more virulent *Plasmodium falciparum* species of malaria, instead of *P. vivax*, the outcomes may have been more tragic.

diff. strains can be more or less fatal

The histories of malaria in Archangel, Bengal, and Palm Beach teach us a second lesson. The convergence of the three conditions that drove these epidemics did not happen by accident. Rather they were strongly influenced by wider societal forces. In Archangel, the Russian Revolution and Civil War produced highly vulnerable human populations and introduced new strains of malaria. In Bengal, the expansion of British colonial rule and commercial agriculture led to the construction of the embankments that altered the flow and distribution of water. In Palm Beach County, the failure of sanitation workers to clear vegetation from drainage canals encouraged vector breeding, while a globalized economy that moves individuals and parasites around the world introduced *P. vivax* malaria. Remove these wider societal forces and the impact of malaria would have been greatly reduced or eliminated in all three histories. Conversely, as we will see in chapter 2, changing global forces could also contribute to the disappearance of malaria.

social/economic

That societal forces have played and continue to play a critical role in the epidemiology of malaria is hardly an original observation. Malariologists have long recognized the social and economic determinants of the disease, and an extensive literature illustrates how patterns of social and economic development have interacted with local ecological conditions to shape its transmission.[12] Yet this interaction, what we might call the human ecology of malaria, has tended to disappear from view in the face of expanding scientific knowledge of the biology of malaria. Since the end of the nineteenth century, efforts to control and eliminate malaria have focused increasingly on attacking mosquitoes or malaria parasites, while largely ignoring the societal forces that contribute to the disease. Although this approach has had some success, it has failed to bring malaria under control, let alone eliminate it as a public health problem within many tropical and subtropical regions of the globe. In order to understand and to respond effectively to the persistence of malaria as a global health problem, it is critical to view the disease as part of a wider historical narrative in which human actions have encouraged the breeding of malaria vectors, exposed populations to infection, and facilitated the movement of malaria parasites.

At the center of this narrative have been efforts to exploit the land through the development and expansion of agricultural production. Other human activities have shaped the epidemiology of malaria; warfare, mining, and the building of roads and railways have all been associated with its appearance or expansion. Population movements, whether voluntary or spawned by the forced recruitment of labor, warfare, or famine, have also contributed in a number of ways to the emergence of malaria in various parts of the globe. Urbanization, particularly in the tropics, has produced conditions that generated malaria transmission. Anopheline mosquitoes have readily exploited the various man-made conditions; they can breed in rain barrels, drainage ditches, discarded canisters, and tires and take advantage of the high concentrations of often poorly housed human populations located in urban settlements. Malaria, or "autumnal fever," so named because it often occurred in late summer and fall in temperate regions of the globe, plagued the populations of Baltimore, Philadelphia, Washington, and New York before adequate sanitation and drainage systems were developed during the nineteenth century. Rome, Paris, and even London were not immune from the disease, and malaria continues to plague a number of cities in Africa and South Asia.

Through the early twentieth century, however, the vast majority of the world's population lived in rural areas and worked the land for a living. During this long period, malaria was primarily a rural disease and closely tied to the history of agricultural production. Even today, as we will see in chapter 7, the opening up of new lands for farming in the Amazon region of Brazil and the cre-

rural vs. urban

ation of agricultural plantations in places like Swaziland and Central America have contributed to the global resurgence of malaria. While it is important not to view agricultural transformations as a monocausal determinant of malaria, in isolation from other social and economic changes, agrarian development has played a central role in the global history of malaria and provides a useful foundation upon which to construct a global narrative of the disease through the middle of the twentieth century. We examine this history in the first four chapters of this book.

AGRARIAN TRANSFORMATIONS AND THE HUMAN ECOLOGY OF MALARIA

Where malaria parasites existed, or were introduced, early efforts to clear the land for cultivation frequently created ecological conditions that fostered malaria transmission. Clearing away forest cover and vegetation, turning up the soil, and irrigating the land created new collections of water and encouraged the breeding of anopheline mosquitoes in close proximity to settled human populations. The extent to which early agricultural communities were subject to malaria, however, depended on a range of local factors, including the proximity of people's homes to their fields (and other housing conditions), the presence of livestock that provided an alternative source of blood meals for local mosquitoes, and the breeding and feeding habits of the local mosquitoes. In fact, if you ask a malariologist how environmental changes associated with the early agriculture might have encouraged malaria transmission, the answer will undoubtedly begin with, "It depends . . ." It is difficult to make generalizations about the human ecology of malaria. Nonetheless, as we will see in the histories of agricultural development in such areas as tropical Africa, Sardinia, South Carolina, and the upper Mississippi Valley, malaria plagued early agricultural settlements over wide areas of the globe.

Although the initial opening up of lands for cultivation was often linked to the spread of malaria, the disease tended to disappear as agricultural communities matured and expanded, primarily for two reasons: populations exposed to recurrent infections developed resistance to malaria over time; and improvements in agriculture eliminated anopheline breeding sites and reduced malaria transmission. The prominent eighteenth-century Philadelphia physician Benjamin Rush, who like many observers of the time viewed malaria as the product not of mosquitoes and parasites but of miasmic gases generated by excessive moisture in the soil and rotting vegetation, nonetheless correctly observed that, "while *clearing* a country makes it sickly . . . *cultivating* a country, that is, draining swamps, destroying weeds, burning brush, and exhaling the unwholesome and superfluous moisture of the earth, by means of frequent crops of grains, grasses, and vegetables of all kinds, renders it healthy."[13] CUltivation

Improvements in agriculture were often associated with other changes that contributed to the disappearance of malaria. These included the acquisition of livestock, improved housing, better diets, and general improvements in overall health. As we will see in chapter 2, the disappearance of malaria over wide areas of western Europe and the United States from the end of eighteenth century was largely due to such improvements. *life improvements*

By contrast, where improvements in agriculture did not occur and methods of cultivation remained largely unchanged, the severity of malaria might diminish, yet the disease would not disappear. This situation describes much of sub-Saharan Africa through the middle of the twentieth century. While some African populations found ways to reduce exposure to malaria, primarily by avoiding areas in which the disease occurred, this was not always possible. Over much of the continent, surviving malaria depended on the development of acquired resistance or genetic adaptation to the disease.

Whether agricultural improvements led to a decline in malaria was dependent on the economic and political conditions under which agricultural systems evolved. Agriculture stagnated where market forces were not sufficient to stimulate improvements in production. Yet it also languished where the structure of commodity markets, and specifically the prices paid to producers for their crops, limited the ability of local producers to accumulate capital. The latter phenomenon, as we will see in chapter 4, was a common occurrence in wide areas of tropical Africa, southern Asia, and Latin America during the nineteenth and twentieth centuries.

Transformations in agricultural production did not always lead to reductions in malaria. They could in fact have the opposite effect. This is precisely what happened in the Roman countryside during the third century BCE and in southeast England in the sixteenth and seventeenth centuries. Agricultural transformations were also partly responsible for the rise of malaria in Bengal in the nineteenth century. The outcome depended on a range of local environmental factors and on the conditions under which the agricultural production was organized.

The health benefits of expanding agricultural production were often limited where the ownership of land and capital became concentrated in the hands of large landowners or corporations at the expense of peasant farmers. Malaria often thrived among poor farmers who were displaced from the land or who retained a small plot of land but lacked the resources to improve it. Such populations often lived in inadequate shelters and subsisted on nutritionally poor diets. In other words, malaria persisted where agricultural systems exposed human populations to infection, or reduced their ability to resist infections. As we will see in chapter 2, the emergence of malaria in the Roman Campagna accompanied the creation of large estates and the dispossession of local peasants. Malaria followed similar patterns of development on tropical and subtropical plantations, as well as on the *latifundia* of southern Italy and the cotton plantations of the American South during the nineteenth and early twentieth centuries.

Dispossession, of course, did not always lead to poverty and disease. Where agricultural expansion coincided with the rise of other industries, as it did in northern Europe and North America in the nineteenth century, dispossessed rural populations could seek employment in towns and cities. While the conditions of early industrial life were far from ideal, industrial employment could provide an escape from rural poverty at the same time that it reduced exposure to malaria. Where industrial employment was not available, or where employment opportunities were limited, or where farmers chose to resist the abandonment of a rural way of life, emigration, domestic or overseas, provided another means of escaping poverty and malaria (although, as we will see in the experience of men and women who sought to escape malaria and poverty in seventeenth-century England by emigrating to America, this reprieve could be short-lived). If none of these options were available, the rural poor could face increasing poverty and a rising burden of malaria.

Finally, even where agricultural developments encouraged the disappearance of malaria, the disease could return. Malaria has always been an opportunistic disease that can quickly take advantage of conditions created by the breakdown of mature agricultural communities, following social, economic, or political disruptions, such as we saw in the cases of Russia and Bengal. *can come*

These three scenarios—stagnating agriculture and continued *back* exposure to malaria, agricultural improvement and a reduction in *any-* malaria, and agricultural expansion and continued or increased *where* malaria—could and did occur over wide areas of the globe. Yet one can see a gradual divergence in the ways in which agriculture and malaria evolved in the temperate North and the tropical and subtropical South from the end of the eighteenth century. Increasingly, the North experienced the capitalization of agricultural production and the decline of malaria, whereas the South suffered agricultural stagnation or forms of production that encouraged the continuation or expansion of malaria. *] v. interesting paragraph*

This divergence was reinforced by inequalities in political and economic power between the North and South. These inequalities began to emerge as early as the sixteenth century, but expanded dramatically during the nineteenth and twentieth centuries with the growth of industrial capitalism in Europe and America, the spread of European colonial rule, and the incorporation of local tropical economies into global markets.[14]

The first half of this book examines how these broader patterns of global historical development shaped the history of malaria through the beginning of the twentieth century. It does this by linking together a series of detailed historical case studies. These case studies cannot hope to cover the diversity of local experiences that make up the global history of malaria and are not intended to provide a comprehensive view of that history. They are instead core samples taken from specific points in the broader historical landscape of malaria in order to explore in some detail the complex ecological processes that have shaped that landscape.

first emerged in subsahan Africa

I begin this story in Africa. Although there are conflicting theories about the origins of human malaria, it is highly probable that the disease first emerged among farming populations living in the tropical forests of sub-Saharan Africa. The early history of malaria in Africa illustrates the importance of the introduction of agriculture for the emergence of this disease. It also provides an opportunity to explore the basic biological characteristics of malaria and its epidemiology. From Africa I turn in chapter 2 to the expansion of malaria northward into Europe and the New World and to its disappearance from these regions from the middle of the eighteenth century. Chapters 3 and 4 examine why the history of malaria in tropical and subtropical regions of Africa, Asia, and Latin America, as well as in the American South and Italy, followed a different path, becoming entrenched and in some areas expanding its hold on local populations during the nineteenth and twentieth centuries. These chapters help explain why malaria is currently a disease largely limited to tropical and subtropical regions of the globe.

Taken together, chapters 1–4 reveal how the historical epidemiology of malaria has been driven by the interplay of malaria parasites, anopheline mosquitoes, and human hosts operating within wider social, political, and economic contexts. They consider the human ecology of malaria and provide important lessons about the epidemiology of the disease. The final four chapters examine why these lessons have been largely forgotten in the wake of human discoveries of the parasite that causes malaria and the role of anopheline mosquitoes in its transmission. They also explore how these discoveries were translated into efforts to control the disease and why these measures have failed to eliminate malaria as a public health problem over wide areas of the globe.

Before we begin to examine the early history of malaria, a few words of warning may be in order. Constructing a global history of malaria presents many challenges. Aside from the difficulty of making categorical statements about a disease that is shaped by so many local contingencies, documentation on the history of malaria is both limited and contested. This is especially true for the earliest periods, where, for example, the value and meaning of ge-

Issues

netic evidence for the tracing the evolutionary history of malaria parasites, as well as the use of DNA evidence for dating the spread of malaria from human remains, have been the subject of considerable debate. There are also large gaps in our present knowledge of the history of malaria in many parts of the globe. In particular, we know very little about the history of malaria over much of Africa, Asia, and Latin America prior to the twentieth century. As future research uncovers the history of malaria in different parts of the globe, the global patterns identified in this study may have to be revisited.

Constructing a global historical narrative for malaria is also difficult because, for much of the period covered in the first four chapters of this study, there was no knowledge of the role of the malaria parasite in the transmission of the disease and thus no basis on which to make a definitive diagnosis of malaria. Without such evidence, we must rely on contemporary descriptions of symptoms to identify the disease. Differences in nomenclature and descriptive evidence, however, often make it difficult to know for certain whether a document is referring to malaria or some other disease associated with fever. The terms *fever* and *ague,* often associated with malaria-type fevers, were used to describe other diseases, including typhoid. In other cases, the symptoms and conditions described are subject to alternative interpretations. For example, in 1780 an epidemic of what was called remittent fever swept through the city of Philadelphia, taking hundreds of lives. The disease resembled outbreaks of epidemic malaria in many respects, and the term *remittent fever* was often associated with malaria-like conditions. However, more recent analyses of the data have led observers to conclude that the epidemic was not malaria but dengue fever. Retrospective diagnosis is an exercise fraught with difficulties.

Generally, however, historians have suspected that reports of fevers and chills that occur every two or three days, during summer and autumn months in temperate and subtropical climates, or following rains in tropical climates, among people living close to marshy environments, are describing in all likelihood symptoms caused by malaria. In addition, reports that fevers were effectively treated with cinchona bark, and from the early nineteenth century, quinine, provide another indication of the malarial nature of the fevers. Both substances have long been recognized to alleviate the symptoms of malaria. As American malariologist M. A. Barber observed in the 1920s, "When large numbers of people come down in the autumn with attacks of chills and fever which often occur every other day and yield to quinine, a microscope is hardly needed to prove that a good deal of real malaria is concerned in the matter."[15]

Despite these methodological problems, I believe it is possible to use the available data to create a historical narrative of what must have occurred, and I leave it to future historians to determine whether this narrative is accurate. In the meantime, this is what I think happened.

do best to describe

CHAPTER ONE

Beginnings

THE EDGE OF THE FOREST

The tropical forests of Africa once stretched in a broad arc across West and Central Africa, extending eastward into the Great Rift Valley and southward into present-day Mozambique, Zimbabwe, and Angola. Today what is left of the forest is limited to an area largely concentrated around the Congo River basin, southern Cameroon, and Gabon, along with a few remaining clusters scattered over West Africa. Climatic change in association with the early development of agriculture in sub-Saharan Africa contributed to the loss of forest cover beginning some 8,000 to 10,000 years ago. It was in the African rain forest and the expanding areas of cultivation cut out of the forest that in all likelihood malaria had its human origins.

Malaria is caused by the invasion of the human bloodstream by a parasitic protozoan of the genus *Plasmodium*.[1] Early forms of this parasite infected the ancestors of the taxonomic order to which mosquitoes belong, the Diptera. This may have occurred as early as half a billion years ago. Within the Diptera the premalaria parasites reproduced sexually, as their descendants continue to do today. Malaria parasites subsequently infected vertebrate animals upon which Diptera and other blood-feeding arthropods fed.

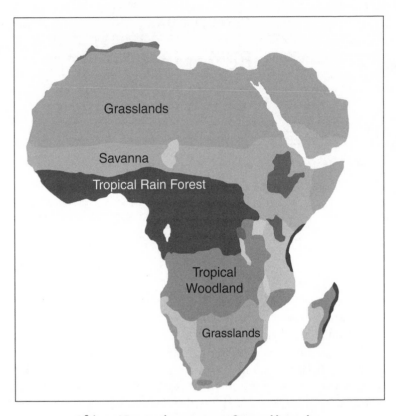

African Vegetation, 7,000–8,000 Years Ago
Source: "Africa during the Last 150 Years," Environmental Sciences Division, Oak Ridge National
Laboratory, www.esd.ornl.gov/projects/qen/nercAFRICA.html.

Within the infected vertebrates, the malaria parasites developed a
second, asexual, reproductive cycle called schizogony.

The asexual reproductive cycle occurs in humans today and
is responsible for the symptoms malaria produces. The cycle be-
gins with the injection of the sporozoite stage of the malaria parasite into the human bloodstream following the bite of an infected
female anopheline mosquito. Only female mosquitoes feed on
vertebrate blood. They do this because they need a high-energy
food to complete the production of eggs and thus the reproduction of their own species. Vertebrate blood provides a convenient
source of this food. The sporozoites quickly enter the host's liver.

There they undergo a nuclear and cellular division into numerous daughter cells, or merozoites. Through this process a single sporozoite may generate from 10,000 to 30,000 merozoites. When mature, usually about 5 to 6 days later, the merozoites enter into the human bloodstream, where they invade the circulating red blood cells, usually, but not always, one to a cell. Within these colonized blood cells, the parasite grows, consuming hemoglobin from the host cell. It then divides again, each parasite producing on average 8 to 16 daughter cells that eventually burst forth, destroying the red blood cell and entering the bloodstream. These new merozoites invade other red blood cells, repeating the process of asexual reproduction. The invasion of new red blood cells repeats itself every two or three days, depending on the species of malaria. This periodicity of blood cell invasion and bursting produces the classic human malaria symptoms of recurrent fevers and chills.

Not all merozoites invade red blood cells. Some are differentiated into male and female forms, known as gametocytes, which circulate in the peripheral bloodstream. Female anopheline mosquitoes will ingest some of these gametocytes during their blood meals. These ingested gametocyte forms mature into sexual forms in the mosquito gut where they mate and eventually make their way through the wall of the host mosquito's gut. Once there, they complete the sexual reproduction stage of the parasites' life cycle, producing fresh sporozoites that invade the mosquito salivary glands and infect new human hosts during the next blood meal.

The development of this form of asexual reproduction was an important evolutionary adaptation, for it permitted the malaria parasite to greatly increase its proliferative potential to levels much higher than would be possible through sexual reproduction alone. The parasite numbers can expand 8- to 16-fold with each cycle of red blood cell invasion. This greatly increases the parasite's chances of survival. At the same time, the production of large numbers of asexual forms of the parasite within human hosts accounts for the often severe disability associated with malaria infection. Finally, the complex reproductive life cycle of malaria parasites means that their survival depended on the existence and close association of vertebrates and mosquitoes.

From the earliest malaria parasites, numerous species of *Plas-*

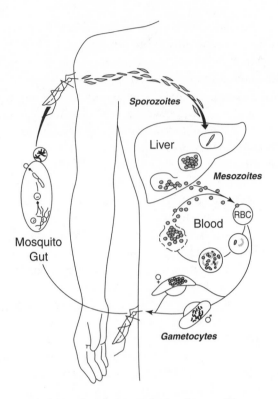

Sporozoites

Liver

Mesozoites

RBC

Blood

Mosquito
Gut

Gametocytes

Life Cycle of the Malaria Plasmodium
Source: Louis H. Miller et al., "Research toward Malaria Vaccines," *Science* 234, 4782 (December 12, 1986): 1350.

modium emerged, each adapted to a specific vertebrate host. Only four of these, however—*Plasmodium malariae, P. ovale, P. vivax,* and *P. falciparum*—currently infect humans. Each of these has a close relative within primates, and genetic analysis suggests that they evolved in Africa within the population of early hominids that diverged from great apes, venturing out into the African savanna some 5 million years ago (though some have argued that *P. vivax* originally infected *Homo* populations in Southeast Asia).[2]

It is difficult to determine just when malaria parasites infected human populations in Africa. *P. malariae, P. vivax,* and *P. ovale* may have infected the earliest hunters and gatherers. These parasites only attack specific red blood cells and therefore do not cause the severe anemia that may lead to the death of their human hosts.

In addition, they are able to exist within humans for long periods without reinfection. *P. vivax* and *P. ovale* do this by producing a resting form known as a hypnozoite during the liver stage of their asexual reproductive cycle. This form can remain dormant in the liver for as long as three to five years before undergoing division and producing merozoites, which invade red blood cells and cause disease symptoms. The ability of these three parasites to survive without reinfection, combined with their limited lethality, allowed them to survive among sparse populations.

By contrast, *P. falciparum* is unlikely to have infected small mobile human populations and probably emerged in humans more recently than the other three malaria parasites. Unlike *P. malariae, P. vivax,* and *P. ovale, P. falciparum* causes severe illness leading not infrequently to the death of its human host. This inhospitable behavior stems from the fact that unlike the other three strains of malaria parasites, which invade selected types of red blood cells, *P. falciparum* invades all types of red blood cells. The number of infected and destroyed cells reaches very high levels, approaching in some cases 80 percent of the total. By contrast *P. vivax* infects no more than 2 percent of red blood cells.[3] Today *P. falciparum* accounts for the vast majority of human deaths from malaria worldwide. *P. falciparum* parasites also survive within human hosts for a relatively short period of time that is measured in months, not years. They either disappear after causing a short acute infection or kill their host. The propensity for *P. falciparum* to kill its human hosts, together with its relatively short survival time, meant that it required large numbers of potential human hosts to survive. *P. falciparum* was probably not able to gain a foothold among African populations until larger sedentary communities emerged in association with the evolution of domestic agriculture in Africa between 4,000 and 10,000 years ago.[4]

Larger, more stable human populations, while critical for the survival of *P. falciparum* parasites in human populations, facilitated the transmission of all malaria parasites. It is unlikely that high levels of transmission were achieved in small hunting and gathering populations. To understand why this is so, we need to stop for a moment and examine just how malaria parasites are transmitted from one human host to another. For transmissions to occur, a malaria parasite must be ingested by a female anopheline mosquito, undergo sexual reproduction within the female *Anopheles*, and then be passed to a second human through the bite of the infected mosquito. Though this may sound simple, it is not. Only the sexual or gametocyte stage of the parasite can infect mosquitoes, and gametocytes appear only sporadically in the circulating blood of an infected human, "coming without warning and disappearing again without apparent reason."[5] Moreover, the more chronic the infection, the scarcer the gametocytes.[6] Thus, an anopheline mosquito may take multiple blood meals from infected humans without becoming infected. M. A. Barber, trying to identify individuals with malaria capable of infecting mosquitoes in Panama, examined 1,500 infected cases and found only 14 with sufficient numbers of gametocytes to serve as infectors, and only 5 of those successfully infected an anopheline mosquito.[7] So the chances that a mosquito would become infected within a sparsely settled nomadic population of hunters and gatherers in Africa were very low. As the size and density of human populations increased, so did the numbers of potential sources of infection.

Transmission is also limited by the fact that the parasite must undergo sexual reproduction within the female mosquito before it can be passed on to another human. The reproductive cycle within the female *Anopheles* takes roughly 14 days, depending on the species of parasite and the ambient temperature.[8] For this to happen the female *Anopheles* must survive long enough for sexual reproduction to occur in the parasites. The average female anopheline mosquito survives from 10 to 21 days, which is barely enough time to complete the reproductive cycle. A study done in Africa estimated that only 10 percent of mosquitoes infected with *P. falciparum* malaria would survive long enough to complete a 14-day reproductive cycle.[9] Higher densities of human hosts, however, increased the likelihood that a female anopheline mosquito would be infected early enough in her life-span to complete the reproductive cycle. Thus, the likelihood of a mosquito being infected with malaria and transmitting it to another human host was affected by the size and mobility of the human population. The odds increased significantly when populations grew in size and became more sedentary.

The combined obstacles to transmission created by the *Anopheles*'s short life-span and the low percentage of infected humans capable of infecting other *Anopheles* may lead one to wonder how so many people become infected with the disease. The answer lies in numbers. Malaria transmission requires high densities of anopheline mosquitoes. The precise numbers depend on the species and its ability to transmit malaria. Highly efficient transmitters, like the African *Anopheles gambiae* mosquito, require relatively small numbers, whereas an inefficient transmitter like *Anopheles albimanus,* found in Central America, require very high numbers. In general, however, even though the odds are low, there are a lot of players. High densities of anopheline mosquitoes enable transmission to occur despite the low probability that any one mosquito will be infected or live long enough to permit the completion of the sexual reproduction of the malaria parasite. This fact explains why changing ecological conditions that increase or decrease the reproductive capacity of local malaria-transmitting mosquitoes play such a critical role in the epidemiology of malaria.

Ecological changes associated with the rise of stable agricultural populations in Africa contributed to the transmission of malaria in humans in one more important way. They played a part in the evolution of the *A. gambiae* mosquito, the most efficient transmitter of malaria in Africa, if not the world. The efficiency of *A. gambiae* in transmitting malaria derives largely from its preference for taking its blood meals almost exclusively from humans and not from other mammals, such as cattle. All other factors being equal, the probability of a malaria infection being transmitted from one human to another is proportional to the fraction of blood meals the mosquito takes from humans. In most parts of the world, the percentage of blood meals that anopheline mosquitoes take from humans is less than 50 percent of the mosquitoes' total blood intake, and often less than 10 to 20 percent.[10] *A. gambiae,* by comparison, take between 80 and 100 percent of their blood meals from humans. *A. gambiae* mosquitoes are also highly susceptible to malaria parasites. This means that in malaria-endemic areas a high percentage of female *A. gambiae* will be infected with malaria, and, in turn, a high percentage of their blood meals will result in the injection of malaria sporozoites.

Because of this combination of susceptibility and preference for human blood, in areas where *A. gambiae* are present and transmission occurs throughout the year, a person may receive up to 1,000 infective bites a year.[11] The efficiency of *A. gambiae* mosquitoes in transmitting malaria—its high vectorial capacity—means that malaria transmission could be maintained with small numbers of *A. gambiae* present. Consequently, malaria control programs based on mosquito elimination in Africa have faced significant problems. Where environmental conditions allowed for the breeding of large numbers of *A. gambiae,* entire human populations could be infected in a short period of time. The evolution of *A. gambiae* mosquitoes contributed greatly to the growing importance of malaria as a source of sickness and death in Africa.

So what explains the extreme human-biting preference of *A. gambiae* mosquitoes? The breeding sites for *A. gambiae* are typically small temporary pools of fresh water with bare edges exposed to sunlight. Such sites occur naturally in savanna areas of Africa, although they are multiplied by activities related to the cultivation of crops, including breaking up the soil and clearing the forest cover. You would expect that the spread of domestic agriculture in the savanna region of West Africa would have created breeding sites close to human settlements, encouraging local anopheline species to adopt human feeding behavior. But within the savanna areas of Africa, domestic agriculture emerged in close association with the keeping of domesticated animals, including cattle,[12] and farmers often kept their cattle in close proximity to their homesteads. The presence of alternative hosts would have limited the selective pressure on *A. gambiae* to develop its characteristic preference for human blood. We need another explanation, and it comes from the early development of forest agriculture.[13]

Within the forest, high levels of rainfall provided ideal breeding grounds for *A. gambiae,* as long as human populations destroyed the thick vegetation cover and the very absorbent layer of humus on the forest floor. This is precisely what occurred with the expansion of forest agriculture. While the earliest forest dwellers relied primarily on hunting and gathering, along with some fishing, for their food supplies, they also developed "vegeculture" (also referred to as protoculture or paracultivation), a food-growing technique based on the recognition that certain roots, fruits, and trees would grow again in the same place if slips were put in the ground and a little weeding was done afterward. The practice eventually led to the domestication of several types of yams, along with calabashes and oil and raffia palms. This pattern of forest exploitation may have existed for thousands of years, and linguistic evidence suggests that the occupation of the African tropical forest dates from around 8,000 years ago.[14]

A second phase of forest agriculture began when forest dwellers came in contact with African farmers growing cereal grains in the margins of the forest edge—between 4,000 and 6,000 years ago. This contact evidently led some forest populations to pay more attention to planting root crops and to adopt some cereals. It also encouraged them to clear away more forest cover by burning and cutting with stone axes (slash-and-burn agriculture) and to establish regular fields. At the same time, the forest dwellers adopted sheep and goats. The size of these forest communities expanded over time and evolved into settled villages surrounded by fields. The final exploitation of the forest occurred following the spread of iron-making technology around 2,000 years ago.

It is difficult to say just when forest dwellers created conditions that encouraged the anthropophilic behavior of *A. gambiae* mosquitoes. In all likelihood, however, it began during the long period in which cultivation was evolving, creating breeding conditions that favored the reproduction of *A. gambiae* mosquitoes, but before the widespread adoption of domestic animals that would have provided alternative vertebrates on which to feed.

The development of *A. gambiae* would have accelerated the transmission of *P. malariae, P. vivax,* and *P. ovale* malaria among forest-dwelling agriculturalists because even a relatively small number of these mosquitoes could infect a large number of human beings. *A. gambiae* may also have contributed to the evolution of *P. falciparum* parasites by increasing the prevalence of multiple infections in human hosts and enhancing competition among different species of malaria parasites. Within this competition, the survival of each species would have depended on its ability to colonize red blood cells. This would have favored the emergence of a fast-growing aggressive species responsible

for acute, short-lived infections in the human host—specifically,
P. falciparum.[15] However, the ecological conditions that would
have given rise to *A. gambiae*'s preference for human blood—forest agriculture without alternative animal hosts—probably occurred in small-scale communities that would have been wiped
out by the aggressive pathogenicity of *P. falciparum.* Only during
the second phase of forest exploitation, when forest communities
grew into settled villages, did forest agricultural populations reach
a size that would have supported *P. falciparum.*[16] It also is possible
that *P. falciparum* did not emerge originally within forest communities but was associated with agricultural activity in the savanna
areas of West Africa in association with the expansion of members
of the *A. gambiae* complex. In either case, although the evolution
of *A. gambiae* may have contributed to the evolution of *P. falciparum,* the two events probably did not occur simultaneously.

The early emergence of malaria in Africa reveals how, from the
time of its initial appearance in human populations, its epidemiology was driven by the interplay of biological and social forces.
Social and economic transformations associated with the early development of agriculture in tropical Africa created new ecological
conditions that encouraged the evolution of more efficient
vectors, gave rise to new more deadly forms of malaria parasites,
and generally expanded malaria transmission. The association between malaria and early efforts to expand agricultural production
occurred, as we will see later, over wide areas of the globe.

The emergence of *P. falciparum* along with the heightened
transmission of earlier forms of malaria must have taken a toll on
early agricultural populations in Africa, although we have no demographic evidence on which to determine its extent. Over time,
however, African populations developed physiological defenses to
various forms of malaria.

Within much of tropical Africa, with its stable conditions of
transmission, high frequencies of inoculation, and the presence
of *A. gambiae,* a person would rarely go for more than a few days
without an infective bite. Under these holoendemic conditions,
individuals developed resistance to the disease-causing blood
stages of the malaria parasite. This resistance did not prevent infection, but by reducing the reproduction of blood stage parasites,

it limited the severity of the disease.[17] This resistance was specific to each parasite species and evolved slowly over time, particularly where individuals were subject to infection by multiple species.[18] Children younger than five, too young to have developed this resistance, remained vulnerable to high levels of infection, causing severe symptoms, whereas older children and adults were better able to limit the development of the disease. Even today, the vast majority of malaria deaths in tropical Africa occurs among these young children. By contrast, in areas where high levels of transmission occurred seasonally, this kind of immunity was much less developed, and malaria symptoms tended to occur across the entire age range of the population. This pattern, known as hyperendemic or mesoendemic malaria, depending on the level of infection within the population, occurred in subtropical areas of eastern and southern Africa. In areas where malaria seldom occurred, such as the highlands of eastern and southern Africa, populations developed no resistance to the disease, so periodic invasions of malaria often caused high levels of morbidity and mortality.[19]

Acquired immunity was not permanent. Extended interruptions in exposure to malaria parasites reduced resistance. An interval of one to two years without reinfection would be enough to make an individual susceptible again to the full impact of a malarial infection. While such interruptions would have been unusual in areas of stable, or continuous, transmission, they could occur during extended periods of drought. Today, people who are brought up in holoendemic areas of Africa but go to Europe or America for their education may suffer serious bouts of malaria upon their return home.

Intensive transmission also gave rise over time to genetic mutations that provided more permanent protection from malaria. These adaptive mutations include the thalassemias, which occur with considerable frequency among people living in formerly malarious regions of southern Europe and the Middle East; glucose-6-phosphate dehydrogenase (G6PD) deficiency; and several genetically transmitted hemoglobin mutations.[20] Within tropical Africa, two types of human genetic adaptation were particularly important: Duffy negative red blood cells, providing nearly complete protection against *P. vivax* malaria; and hemoglobin S (the basis of sickle cell anemia), which reduces the severity of *P. falciparum* malaria infections.

In people infected with *P. vivax* malaria, the blood stage merozoites enter the red blood cells through a receptor on the surface of the cell known as the Duffy antigen. Expression of this antigen is determined by the presence or absence of two positive forms, or alleles, of the gene that controls this antigen. Most human populations possess two positive alleles for the Duffy antigen. Throughout West and Central Africa, however, nearly 97 percent of the populations possess two negative forms of the gene, and the Duffy antigen is absent. This absence prevents *P. vivax* merozoites from penetrating the human host's red blood cells and thus protects the host from vivax infection. The peoples of this region of Africa are, accordingly, almost completely resistant to vivax malaria. It probably took tens of thousands of years for Africans to achieve this state of nearly complete negativity, because full protection requires the individual to have two negative alleles. Microbiologists date the emergence of Duffy negativity to the end of a very long period of human adaptation dating from between 97,200 and 6,500 years ago.[21]

Whether this evolutionary adaptation was a selective response to a long history of intense exposure to vivax malaria or evolved independently of malaria is still debated. Those who argue that vivax malaria originated in Southeast Asia suggest that Duffy negativity evolved independently of *P. vivax* and prevented it from penetrating large areas of Africa. They note that *P. vivax* creates a relatively mild infection in humans; there would therefore have been little evolutionary pressure on human populations to develop genetic adaptations in response to the disease.[22] On the other hand, those who believe that *P. vivax* originated in Africa argue that Duffy negativity represents a successful genetic adaptation to intense exposure to the parasite. Because the parasite may have been a more serious challenge to human health during earlier periods when it caused infrequent infections, it could have exerted more selective pressure on the exposed human population. Researchers agree, however, that the presence of Duffy negativity provided complete protection against vivax infections for human populations living over wide areas of sub-Saharan Africa, as well as for many of their descendants brought to the New World during the Atlantic slave trade.

There is little doubt that the hemoglobin S mutation evolved in African populations in response to the presence of *P. falciparum*. The mutation provides protection against *P. falciparum* in individuals who inherit a single sickle cell allele, greatly reducing the risk of death from this form of malaria. However, the mutation causes a rare, often fatal, form of anemia in those who inherit both alleles. Because this condition significantly reduces an individual's chance of survival, the sickle cell trait could have persisted in human populations only where the mortality advantage provided by reducing the severity of *P. falciparum* malaria outweighed the disadvantage created by sickle cell anemia. Hemoglobin S is, accordingly, found only in populations who live in areas where *P. falciparum* is, or has been, highly endemic; or in populations, such as African Americans, whose ancestors lived in such areas. In some parts of central Africa, up to 30 percent of the population exhibits this trait, providing the people with protection from *P. falciparum*. The hemoglobin S trait would have taken less time than Duffy negativity to evolve, because *P. falciparum* causes intense selective pressure, and because the protective state requires the development of only one hemoglobin S allele. Genetic analyses date the development of the human protective alleles that cause the sickle cell trait and related *thalassemias* to somewhere between 8,000 and 3,000 years ago, roughly the period of early forest agriculture when *P. falciparum* appears to have emerged in human populations in Africa.[23]

OUT OF AFRICA

While the expansion of malaria in Africa was accelerated by the spread of domestic agriculture, its movement beyond Africa may have preceded this development, dating from the period of the earliest human expansion out of Africa. Movements into the Eurasian landmass may well have been stimulated by the simple ecological reality that the African savanna could not support a population of hunters and gatherers that exceeded two persons per square mile. Even a slow increase in human population would have necessitated an expansion of the limits of the land areas available for exploitation. And the rate of movement must have been extremely slow. Yet, as anthropologist Richard Leakey calculated, populations of hunter-gatherers moving as little as 10 to 30 miles in a generation could have traveled the distance from the Nairobi to Peking in 15,000 years.[24] Obviously such a beeline movement is unlikely, and it took much longer for human populations to expand into the various parts of the globe. Recent genetic research suggests that the initial movement of *Homo erectus* out of Africa began around 1.7 million years ago and that there were at least two later migrations involving later human ancestors[25] This research also indicates that genetic sharing among geographically distinct human populations, combined with additions of new populations emerging from Africa, led to the further evolution of human populations in different areas of the world and gave rise to modern human populations. By 15,000 years ago, modern humans had colonized much of the earth's surface and were subsisting in small groups that hunted and gathered food.

P. malariae, which, as noted earlier, can survive for long periods of time within small populations with limited opportunities for transmission, may have moved out of Africa with the earliest of these small groups of hunters and gatherers, occupying the warmer regions of Europe and Asia. It is also possible that *P. malariae* spread northward into the temperate areas of the Eurasian continent at an early date, because its longevity in humans would also have allowed it to survive seasonal drop-offs in transmission linked to colder winter temperatures. In other words, the traits that permitted it to exist among small mobile populations in tropical Africa made it adaptable to colder climates. The spread of *P. malariae* into continental Europe, however, is unlikely to have taken place before the end of the last glacial period around 10,000 years ago, because low temperatures would have prevented the completion of the sexual reproduction of the parasite within local mosquito vectors. Such temperatures would also have restricted the growth of populations of the anopheline mosquitoes needed to transmit the disease.[26]

Cold temperatures associated with glacial periods probably also prevented *P. malariae* from spreading into North America before the end of the last ice age. In fact, it probably did not enter the

Americas before roughly 1000 CE. This is because the land bridge across which humans migrated between Asia and North America resulted from glaciation and the dropping of the global sea level by as much as 400 feet. Drops in temperature during this period make it unlikely that malaria parasites would have survived the long migratory process that led to the early settlement of the New World.[27] The warming of the earth's climate, following the last ice age, made it possible for malaria to exist during summer months in the temperate climate of Asia and North America. However, melting ice and rising ocean levels eliminated the land bridge between Asia and North America, preventing further human migration and *P. malariae* from entering North America until human contacts between the Old World and the New World began again between 500 and 1,000 years ago.[28]

While the African origin of *P. vivax* remains a source of considerable debate, this parasite shared some of the adaptive advantages of *P. malariae*, permitting it to expand into temperate climates. A latent stage of the vivax parasite (hypnozoites), which resides in the liver for months or even years before entering the bloodstream and invading red blood cells, would have allowed the parasite to survive seasonal drop-offs in transmission associated with cooler climates.

Russian researchers who studied various subspecies of *P. vivax* distributed from Southeast Asia northward into China found that as one moves further north into more temperate climates, the asexual reproductive cycle of *P. vivax* changes. Specifically, the percentage of sporozoites that form hypnozoites increases, rising to 100 percent in species found in temperate areas of China. This research suggests that *P. vivax* adapted its asexual reproductive behavior to the movement of human populations into more temperate climates.[29] One recent estimate of when this happened, based on the time it would take for a population to develop RBC Duffy negative red blood cells, puts the dispersion of *P. vivax* at not earlier than 10,000 to 30,000 years ago.[30] The ability to survive in relatively small populations and in temperate climates made *P. vivax* an early candidate for northward expansion.

The last human malaria parasite to venture beyond Africa was *P. falciparum*. A late bloomer in Africa, *P. falciparum* is unlikely to have expanded out of Africa earlier than 4,000 to 6,000 years ago.[31] Given its inability to survive within human hosts for extended periods, moreover, *P. falciparum* was not well adapted to survival in areas where cooler winter temperatures interrupted transmission. In addition, it required higher average temperatures to complete its sexual reproduction within anopheline mosquitoes. For both reasons, the northern expansion of *P. falciparum*, unlike that of *P. vivax* and *P. malariae*, was limited to the southern tropical and subtropical areas of the Eurasian continent and the Americas. This limitation has continued up to recent times. Finally, the lethality of *falciparum* meant that its northward movement required the growth of agricultural communities large enough to sustain its transmission without killing off all potential human hosts. The last of the four human parasites, *P. ovale*, has a very limited distribution outside of Africa.

Keep in mind that malaria parasites could penetrate only those areas in which local anopheline species were susceptible to infection. Thus, Bruce-Chwatt and de Zulueta have argued that the early penetration of *P. falciparum* into southern Europe was delayed because the available species of *Anopheles* were initially resistant, or refractory, to *P. falciparum* infection. Even the spread of *P. vivax* and *P. malariae* into southern Europe may have been limited prior the introduction of *A. labranchiae* and *A. sacharovi*—mosquito species indigenous to North Africa and the eastern Mediterranean, respectively—sometime in the first millennium BCE. These species are primarily human-biters and are much more efficient transmitters of malaria than preexisting species.[32] The introduction of these species, in turn, depended on the creation of favorable breeding environments produced by soil erosion and coastal flooding, both resulting from cultivation and deforestation.

The expansion of malaria outside of Africa and into the tropical regions of Asia occurred over several millennia, as the parasite spread from one community to another along with the expansion of human settlements, agriculture, and commerce. As in Africa, the initial introduction of malaria, and particularly *P. falciparum*, into tropical communities must have initially limited population growth. Over time, however, Asian populations living in regions of intense transmission, similar to those in tropical regions of sub-Saharan Africa, developed immunities to the disease, restricting its impact on adults and, to a lesser extent, young children. However, the percentage of populations currently exhibiting genetic forms of immunity to various species of malaria, such as hemoglobin S and Duffy negativity, are lower in most of the world than in Africa, reflecting either the later arrival of malaria parasites or lower levels of transmission.

With the data currently available, we are unable to say just when malaria became a problem in various tropical regions of Asia. There are descriptions of malaria-like fevers in ancient medical texts from South Asia and China. Vedic writings from South Asia dating from 1500 to 800 BCE refer to "autumnal fevers" and describe enlarged spleens.[33] Classical Chinese writings contain references to "intermittent fevers," which were seasonal. They also distinguish between fevers by frequency of reoccurrence—either every third day (tertian) or every fourth day (quartan)—from roughly the seventh century BCE. Whether these references are to what we now know as malaria or whether malaria predates the development of these references is impossible to know. We also know relatively little about the precise combinations of social, economic, and ecological forces that contributed to the tropical expansion of malaria. A great deal of work needs to be done on the early history of malaria in this part of the world.

The expansion of malaria into the more temperate regions of the globe probably did not occur to any great extent until the first millennium BCE. This expansion required the growth of local populations and the creation of environments suitable for the breeding of malarial vectors. These two conditions depended in turn on the expansion of domestic agriculture. Once this occurred, malaria began to penetrate various parts of Europe, Central Asia, and eventually the Americas. Malaria also expanded into the temperate regions of southern Africa. While there remain gaps in our knowledge, we know a good deal more about the details of the history of malaria's later northward expansion than we do about the history of malaria's early spread through the tropics. What we know suggests that, just as in Africa, the northern expansion of the disease was tied to the history of agricultural production.

CHAPTER TWO

Malaria Moves North

❖ ❖ ❖

SOUTHERN EUROPE

In the summer of 1623, Pope Gregory XV fell gravely ill. His head ached, his spleen was swollen, and his body shook from alternating bouts of fever and chills. Within days he succumbed. His death triggered a ritualized series of events that led to the election of his successor. Cardinals from across Europe were assembled and sealed into the Sistine Chapel, where they would select the new pontiff. The conclave dragged on for weeks, and one by one the cardinals and their attendants succumbed to the same disease that had taken Pope Gregory. By the time a new pope was elected, six of the cardinals and forty of their attendants had died. The heads of the Catholic Church who died of fever during the summer of 1623 joined a long list of earlier popes, emperors, and other dignitaries who, together with an untold number of common people, had shared a similar fate over the previous centuries. They all died from *mal' aria*, literally "bad air," which was believed to be caused by a noxious gas emanating from swamps and rotting vegetation during the summer and autumn months. *Mal' aria*, or what we now believe to have been malaria, hung over the streets of Rome and the surrounding Roman countryside for centuries.[1]

Just when malaria began to plague the populations of Italy and the rest of Europe is unclear. Sources from as early as the

Why Europe?

sixth century BCE have identified the occurrence of diseases that resemble malaria over wide areas of southern Europe. Hippocratic texts from fifth-century Greece make frequent references to fevers that reoccurred every third (tertian) and fourth (quartan) day during the summer and autumn and involved enlargement of the spleen. Falciparum malaria may have contributed to the Athenian defeat at the battle of Syracuse in Sicily in 413 BCE. According to Grmek, the Syracusan generals restricted the Athenian army to lowland marshy areas known to be afflicted by fever and let the disease do its work.[2] Malaria also became prevalent in Sardinia following the Carthaginian invasion of 502 BCE. By the third century BCE *mal' aria* was taking its annual deadly toll of the inhabitants of Rome and the surrounding countryside.

Over the next two millennia, the disease spread gradually northward across wide areas of western and central Europe, reaching the southeast coastal counties of England by the sixteenth century. From Europe, malaria crossed the Atlantic, probably traveling initially with European colonists and later on slave ships from Africa. The disease found a new home in the emerging agricultural colonies of the New World during the seventeenth century.

The early spread of malaria into Europe and the Americas was closely tied to the changing fortunes of agriculture. In some areas, *link betw. Mal. agri.,* such as on the island of Sardinia, or the early agricultural colony of South Carolina, or the upper Mississippi Valley, it was the initial opening up of the land for cultivation that transformed the region's ecology and created opportunities for the expansion of malaria. In others places, such as the Roman Campagna and the southeast of England, changes in the conditions of agricultural production—the growth of large estates around Rome and efforts to drain coastal lowlands in England—altered the distribution and flow of water over the land, creating opportunities for anopheline mosquitoes and malaria transmission that had not existed before.

By the eighteenth century, improved methods of land management associated with the capitalization of agriculture altered farming practices and began to push the disease out of northern Europe. A similar process began in the upper Mississippi Valley at the end of the nineteenth century. By the beginning of the twentieth century, malaria had largely disappeared over large areas of both continents. The withdrawal of malaria, like its earlier expansion, was encouraged by agricultural transformations that altered the human ecology of the disease.

To understand the history of malaria's northern expansion and withdrawal, and how it was shaped by changing patterns of agricultural production, we need to examine the changing human ecology of malaria in several regions of the North. These cases appear to be representative of the ecological processes that shaped the rise and fall of malaria in Europe and North America. Yet, as noted in the introduction, we need to keep in mind that the historical record of malaria in the North remains incomplete. Further research may uncover ecological patterns that will complicate the story told by the cases presented here.

Sardinia

The island of Sardinia, which rises out of the Mediterranean Sea off the western coast of Italy, appears to have been one of the earliest places where malaria gained a foothold in southern Europe. Being linked commercially to both Europe and North Africa, moreover, Sardinia may have been a conduit by which malaria spread across the Mediterranean and into the temperate regions of Europe, though other strains of malaria may have entered southern Europe from the eastern Mediterranean and Greece. Sardinia is thus an appropriate place to begin our examination of the expansion of malaria into the temperate regions of the globe.

Sardinia's landscape is a mixture of lowland valleys and coastal plains that give way to rocky highland regions that cover most of the island. From Roman times through the first half of the twentieth century, the lowland areas of Sardinia were known as some of the most malarious regions of Europe. Nevertheless, the island was once the site of a healthy and thriving civilization. Traveling northward from the port city of Cagliari, through the Campidano Valley, one comes across the remains of numerous stone ruins, each dominated by a central conical tower or towers, known as *nuraghi*. There are, in fact, an enormous number of these constructions, more than 6,500, scattered over the island.[3] The nuraghi are what is left of a civilization that began around 1800 BCE and lasted until the Carthaginian invasion of the island in the fifth century

Italy

BCE. The highest density of nuraghic settlements is concentrated in the Campidano and Logudoro lowlands, areas that experienced the highest levels of malaria in Sardinia during the early twentieth century. A malaria survey conducted in the 1930s found that Sardinian townships with a high density of nuraghic settlements had an average malaria morbidity rate of 62.8 percent of the total population. One can assume that had malaria existed in these areas to the same degree before 500 BCE, the areas could not have supported the nuraghic civilization that evolved there. The question then becomes, How did Sardinia become one of the most malarious places in southern Europe by Roman times? While historical records for the early history of Sardinia are limited, anthropologist Peter Brown has combined archaeological and biological data to reconstruct the early history of malaria in Sardinia and present a tentative answer to this question.[4]

Farmers, herders

The nuraghic peoples were primarily forager-pastoralists who herded cattle and practiced only a limited amount of agriculture. The islanders did have access to mineral resources and evidently traded with peoples from other parts of the Mediterranean basin. We know that the Phoenicians established trading stations on the coast of Sardinia in the seventh century BCE.[5] This pattern of subsistence did not create environments that encouraged the breeding of local anopheline mosquitoes. Moreover, local *Anopheles* may have preferred to feed on the nuraghic herds rather than humans.[6]

The Carthaginian invasion of Sardinia in 502 BCE dramatically transformed the island's ecology. While some of the nuraghic population remained in the lowlands, most of the people appear to have retreated to the highlands with their herds. There they continued their pastoral existence and resisted further attempts to subdue them. The Carthaginians for their part occupied the main areas of nuraghic settlement in the lowlands, slowly transforming these areas into agricultural lands, establishing large plantations for wheat and flax production. To farm these lands, the Carthaginians were forced to rely on slave labor imported from North Africa, because the nuraghic people resisted efforts to incorporate them into the island's emerging agricultural economy.

The agricultural transformation of Sardinia had several consequences for the history of malaria. First, the introduction of extensive grain cultivation resulted in the removal of the island's forests. This deforestation increased flooding and swamp formation in lowland areas, particularly in the southern parts of the island. These changes created breeding sites for the *A. labranchiae* mosquito, which is both an efficient malaria vector and susceptible to *P. falciparum* malaria. The record of mortality associated with malaria in Sardinia during Carthaginian and Roman times strongly implicates the presence of *P. falciparum,* which may have been introduced by the slaves brought from North Africa.

[handwritten: played a major role in pop. regulation]

The Romans, who occupied Sardinia in 283 BCE, continued the pattern of agricultural exploitation established by the Carthaginians, and Sardinia became a main source of wheat production for Rome. By this time, however, malaria was so extensive that the Romans faced labor supply problems, forcing them to continually replenish the island's work force with slaves, prisoners, and political exiles drawn from various corners of the Roman world. For slaves imported from areas where malaria was not endemic, being sent to Sardinia was tantamount to a death sentence. Over time, however, a portion of the agricultural laboring population of Sardinia evidently developed resistance to malaria. Thus, today, the lowland population of Sardinia possesses high frequencies of thalassemia and G6PD deficiencies—genetic mutations, which, like hemoglobin S mutation, provide protection from *P. falciparum* malaria infections. By contrast, the population of the highlands possesses significantly lower frequencies of these genetic traits.[7]

Sardinia provides an early example of how the expansion of agriculture produced ecological changes that facilitated the emergence of malaria as a serious health problem. Early cultivation and deforestation changed the distribution and flow of water over the land and created anopheline breeding grounds, while the importation of slaves to farm the land introduced new malaria parasites. Malaria continued to plague Sardinia until the late 1940s and early 1950s, when efforts to eradicate local malaria vectors led to the disappearance of the disease from the island.

The Roman Campagna

The early history of malaria in the region of Rome provides a second example of how changing patterns of land use shaped the expansion of malaria into southern Europe. Malaria may have plagued the earliest agricultural communities surrounding Rome, but reference to malarial symptoms in the historical records of Rome is limited before the fourth century BCE. Quite to the contrary, there is compelling evidence that the region supported a thriving agricultural civilization during Etruscan times. The ruins of Etruscan cities together with tombs and Etruscan sepulchers provide evidence of how densely populated the region was in former times. Further evidence is provided by the immense necropo-

[handwritten: some cases, no malaria, but lots of agriculture]

lis and by the impressive ruins that still remain in the vicinity of Rome.[8]

Malaria may have been absent during Etruscan times or it may have been brought under control.[9] Archaeological excavations reveal the existence of an extensive water drainage system consisting of a network of tunnels, or *cuniculi,* constructed to collect rainwater running off hillsides. Although it was not created to prevent malaria, the system may have done so by preventing the collection of water in low-lying areas and thus the creation of breeding sites for anopheline mosquitoes that could transmit the disease.[10]

By the third century BCE, however, the population of the region began to change. Over the course of the late republic and early empire (from 300 BCE to 100 CE), the population of peasant farmers who had cultivated the region emigrated from the agricultural areas surrounding Rome. During the same period, large estates, or *latifundia,* replaced small farms. The process continued during the first centuries CE. Archaeological surveys indicate that the number of occupied sites declined by 80 percent between the first and fifth centuries CE.[11] These transformations were paralleled by a dramatic increase in the intensity of malaria transmission. But was malaria a cause or effect of these demographic changes?

Angelo Celli and Robert Sallares have argued that malaria drove poor peasant farmers to seek their livelihood elsewhere.[12] As they retreated, their farms were bought up by absentee landlords, who then imported slaves to work the land. There may be some truth to this scenario. However, it still does not answer why malaria became a serious health problem during the late republic, when it had not been during earlier times.

Much of the literature that deals with the economic development of the Roman Campagna during the late republic and early empire suggests that the economic transformations just described were triggered not by malaria but by the extensive recruitment of the region's peasant farmers into the Roman army during the Punic and Hannibalic wars. Forced to leave their smallholdings for extended periods of time, farmers conscripted into the army were unable to maintain their farms. As production deteriorated, the absent farmers' families were forced to incur large debts to survive. Many soldiers returned to farms that were ruined by lack of cultivation; others found that their farms had been sold off to cover debts. Liberal land policies following the wars, moreover, allowed wealthy Romans to buy up farms and gain control over public lands.[13] The process of consolidation was further encouraged by the massive importation of grain from newly conquered lands in Sardinia and Egypt. These imports lowered grain prices and undermined the livelihood of local farmers who grew grain for a living. Declining grain prices also led the owners of the newly acquired estates to shift from grain production to cattle raising. Those who engaged in farming focused on wine and olive production. Both activities required capital investments, which prevented the participation of small farmers. In addition, both activities were centered on the hillsides. As a result of these shifts in land use, many of the low-lying areas of the Campagna became fallow or grazing lands. Finally, the large estates took advantage of the plentiful supply of slaves that flooded the Roman market following the wars to provide the labor they needed to run their estates. This labor was both cheap and, unlike the farmers whose lands had been appropriated, exempt from conscription into the army.

Driven off the land, and unable to find employment on large estates, many former peasant farmers abandoned the Roman countryside, seeking employment in Rome or moving to other regions in hopes of finding land to farm. This process of dispossession was uneven, with small farmers holding out longer in some areas than others. But, overall, the process altered the demographic and economic contours of the region. It also opened the land to malaria.

Abandoned or poorly managed low-lying farmlands allowed the collection of water and formation of marshy conditions that provided breeding sites for anopheline mosquitoes. This process was exacerbated by the deterioration of the ancient drainage systems created by the Etruscans and the deforestation of the hillsides that bordered the Campagna.

The process of deforestation extended over several centuries, with the removal of forests accelerating during the third century BCE. The population of the city of Rome grew from about 200,000 to about a million during the last two centuries BCE, producing increased demand for timber to construct houses.[14] In addition, the Punic wars contributed to the demand for timber. During the five years between 262 and 257 BCE, Rome built 200 ships for its navy. Wood was also needed for firewood for smelting metal for weapons. Deforestation led to rapid runoff, choking streams and leading to recurrent flooding of the lowland areas of the region. The resulting swampy conditions provided additional breeding grounds for anopheline mosquitoes. [15]

when did the mosquitos get the parasite

How this expanding population of anopheline mosquitoes became infected with malaria parasites is unclear. However, the extensive use of slaves drawn from all parts of the empire, including Africa and the eastern Mediterranean, certainly provided a potential source of infection.

Finally there is evidence that the climate of ancient Rome became warmer beginning around 300–200 BCE. The average temperature rose by .05 to 2 degrees centigrade. This could have facilitated the reproduction of malaria parasites as well as anopheline mosquitoes.[16]

war, politics

We can see that the interplay of war, politics, and changing patterns of land use transformed the ecology of the Roman Campagna—altering the distribution of water, creating anopheline breeding grounds, introducing malaria parasites, and exposing local populations to malarial infections. As a consequence of these changes, the Roman countryside became highly malarious. The first victims of the rising tide of fever were in all likelihood the farmers who had resisted dispossession. But malaria soon eliminated these hangers-on, killing or driving them out. What land dispossession had begun, malaria finished.

While there appears to have been subsequent periods when the threat of malaria receded, the disease would always return, and it remained a problem until the middle of the twentieth century. It would not be the last time, moreover, that warfare, the dispossession of smallholders, and the disruption of agriculture would give rise to malaria, both here and elsewhere around the globe.

NORTHERN EUROPE

The expansion of malaria into northern and central Europe probably resulted from similar patterns of ecological change associated with human efforts to transform the land. While growing commerce and communications between southern Europe, where malaria had been present from at least the fourth century BCE, and northern and central Europe provided the means by which malaria parasites expanded northward, the extension of agriculture into low-lying, well-watered areas provided opportunities for anopheline mosquitoes to breed. Cooler temperatures prevented the reproduction of the more lethal falciparum malaria in northern and central Europe, but *P. vivax* and *P. malariae* could and did survive and reproduce as far north as England and northern Russia. more history

Malaria appears to have been present in Arab-ruled Spain from at least the eleventh century CE and may have increased in prevalence with the introduction of rice cultivation. In the fourteenth century the crown of Aragon prohibited rice cultivation to prevent disease. Along the Mediterranean coast of France, there is little evidence of malaria during Roman times. Nabaronne was considered a prosperous city, surrounded by fertile lands. However, the cutting down of forests in order to expand cultivation and grazing lands during the thirteenth century converted coastal salt lakes into marshes. In addition, increased runoff from denuded hills changed the course of the Aude River and led to increased flooding of the coastal plains. Together, these changes led to frequent outbreaks of fever, and the population of Nabaronne declined along with the town's prosperity. Similarly, the Loire Valley in France was apparently healthy in the fifteenth century but became fever ridden in the sixteenth and seventeenth centuries.[17] The history of malaria in England beginning in the seventeenth century provides a more detailed example of how malaria may have emerged in parts of northern Europe and how its emergence was linked to agricultural transformations.

Land Reclamation and Malaria in Southeastern England

In the autumn of 1685 Samuel Jeake made the following entries in his diary:

> Sept 9 1685: About noon exactly or very neer, I was taken with a Tertian ague; cold bot not Shaking; after it lay in my head from 6h p.m. till next morning, & I sweat very much in the night. NB

when I was first seized Saturn was just past the Cusp of Medium Coeli & Mars partially opposing the Cusp of the 8th.

Sept 11 1685: About 11h a.m. A 2d fit, cold at first & vomiting, the vomit exceeding sowre, after an hot fit & sweat much in the night.

Sept 13 1685: A 3d fit, cold & vomiting but not so bad as the 2d fit: afterwards hot & sweat in the night. This was the last fit though I were indisposed on the well daies, & after I had lost the ague for neer a week troubled with the spleen, costiveness & a sore mouth.[18]

Jeake was a merchant and dabbler in astrology who lived in Rye near the southeast coast of England. The entries in his diary reveal what was an all-too-common experience during the seventeenth century: episodes of chills and fever, occurring every other day accompanied by vomiting and severe head aches and a "troubled" spleen. Jeake himself suffered multiple occurrences of this illness.

We cannot know for sure from what Jeake suffered in September 1685. He referred to it as "Tertian ague." Yet his symptoms, occurring when and where they did, strongly point to vivax malaria.

Southeastern England, like Archangel, Russia, is not the kind of place that comes to mind when we think of malaria. Yet, during the seventeenth century, there is strong evidence that malaria was a major source of sickness and death among communities living along its coasts and estuaries, especially in the Fenlands, and near the estuarine and marshland coastal areas of northern England. If one drives through coastal counties of southeast England today, it is not difficult to imagine the conditions that gave rise to malaria. While large areas of the contemporary landscape have been reclaimed and converted into expansive farmlands, separated by ditches and canals, vestiges of the marshland environment that once covered wide areas of the region can still be seen in the English Fens and along the coastline in Sussex and Essex. Standing on the stone ruins of the Ypres Tower in the town of Rye, where Samuel Jeake once lived, one can see the salt marshes stretching out toward the English Channel. The coastal marshlands provided

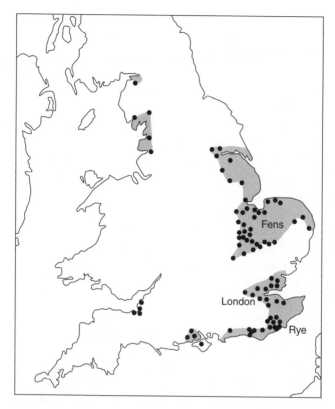

Geographical Distribution of Malaria in England in the 1860s
Source: M. J. Dobson, *Contours of Death and Disease in Early Modern England* (Cambridge: Cambridge University Press, 1997), 348.

both a living for local inhabitants and breeding sites for malaria-transmitting anopheline mosquitoes. Yet this malaria-friendly environment was not its "natural" condition. Like the drained fields that cover much of the landscape today, the anopheline breeding grounds of southeast England and the Fens were shaped by human efforts to transform the region's environment.

Evidence of malaria in the historical records for southeast England is sparse before the middle of the sixteenth century.[19] To the contrary, early parish records indicate that the marshlands and coastal areas of southeast England were relatively healthy during the late medieval period, supporting extraordinary population densities during the thirteenth and early fourteenth centuries.[20]

The silt-rich Fens, south of Lincolnshire, were crammed with people, and in Essex there was a thickly settled belt in the central and northern parts of the county extending south from East Anglia. Taxpayer densities in this area frequently exceeded 20 per square kilometer.[21] However, population densities declined dramatically, as they did elsewhere in England, from the second half of the fourteenth century through the fifteenth century, as a result of recurrent epidemics of plague. In contrast to other parts of Europe, population recovery was delayed. Thus taxpayer densities in the early sixteenth century in most of Essex were less than 6 per square kilometer.

The total population of England rebounded from the middle of the sixteenth century through the end of the seventeenth century, rising from an estimated 2.4 million to just over 5.6 million people, as plague and other infectious disease declined and fertility increased. The population expansion of the sixteenth and seventeenth centuries had important consequences for the social and economic development of English society.

As more people filled the land, the portion of the population without sufficient land to support a family, and thus dependent wholly or partially on wages, increased. There was, in addition, a growing mobility, with people in search of employment moving from overcrowded districts, or districts with inadequate resources, to ones less developed, as well as to cities. The population of London grew from approximately 60,000 people in the 1520s to 200,000 by 1600, and doubled again to 400,000 by the 1650s. There was also during the seventeenth century an increase in emigration to Ireland and the Americas.

The overall increase in population, together with the growth of cities, contributed to a rising demand for food, and a series of bad harvests at the end of the sixteenth and beginning of the seventeenth century reinforced this demand. Growing demands for food led farmers to expand production. It also encouraged the commercialization of agriculture and a movement away from the largely subsistence-based farming economy that had existed at the end of the fourteenth century. Increasingly, agricultural production was shifting from smallholder farming to larger-scale farms run by yeoman farmers. Lawyers, bankers, and merchants also invested their capital in farmland. Finally, the nobility sought to expand the productivity of their large estates by increasingly converting forest and other common lands to arable farming or grazing lands.[22]

In the marshlands of southeast England, and in the Fenlands of Cambridgeshire and Lincolnshire, the rising demand for farmland and grazing lands led to efforts to drain the marshes. These efforts began in a somewhat piecemeal fashion during the second half of the sixteenth century and continued with renewed intensity during the first half of the seventeenth century. Early efforts to drain the land created conditions that fostered the introduction of malaria, while later efforts contributed to the disappearance of the disease.

Agricultural production in the coastal regions of southeast England had for centuries been hampered by the incursion of coastal tides, which flooded fields with salt water.[23] Inhabitants of the marshlands had tried to hold back the tides in order to win back land from the sea and establish pastureland for cattle and sheep or arable land for tillage, while ridding the lands of their high salt content. More concerted efforts followed the passage of the General Drainage Act of 1600. This act made it possible for large landowners and investors to initiate drainage schemes that overruled existing common rights. During the early seventeenth century Dutch immigrants introduced land reclamation methods that had been successfully employed in Holland. This led to the building of embankments to wall off the sea, and the construction of new rivers to drain the land. Drainage efforts increasingly attracted capitalist adventurers, known as "undertakers" or "projectors," who initiated drainage projects in return for a portion of the reclaimed land for their own use.

While those supporting the drainage projects viewed them as part of a process of modernization that would lead to economic betterment, many of the poorer inhabitants of the Fenlands viewed these lands as commons and regarded the projects as a new form of enclosure and a threat to their way of life. These poorer inhabitants used the Fens for gaming and fishing and keeping livestock. They also grew crops there. In short, their livelihood, poor as it was, depended on their continued exploitation of the Fens and marshes. As a result, they actively resisted the drainage projects. In the Fenlands, this resistance led to riots in the 1630s, with protestors destroying drainage systems and taking armed control of the flood-control gates. These gates were turned to their own purposes, inundating the reclaimed fields and keeping them from draining.[24] Drainage efforts came to a standstill during the Civil War but quickly regained momentum following the war with the support of the Commonwealth government. Yet the success of the drainage projects was less than what their promoters had envisioned.

lots of environ-mental agri change

The various drainage schemes initiated up through the middle of the seventeenth century opened up new lands for pasturage and farming, while creating pools of stagnant water and marshes. In the southeast counties, embankments designed to hold back the sea altered the distribution and flow of water over the land by preventing water from draining back to the sea. This led to the development of brackish pools and swampy areas that provided ideal breeding places for anopheline mosquitoes. The species probably included *A. atroparvus,* which breeds well in these environments and which, in more recent times, was the primary vector for malaria in this part of England. In addition, the draining of some land led to the inundation of neighboring fields. In the Fens drainage efforts in areas that were covered with peat not only dried out the land but also caused it to sink. In some cases, the land sank beneath the level of the rivers into which they were supposed to drain, making it difficult to keep them from flooding again. Moreover, because the peat areas were located in the interior of the Fens, these areas sunk below those closer to the sea, inhibiting the drainage of water that flowed into the Fens from surrounding higher areas. These problems not only prevented the hoped for economic progress, they often made matters worse and contributed to the further expansion of malaria.[25]

Malaria became a significant cause of mortality in the marsh parishes during the sixteenth and seventeenth centuries. Descriptions of the disease abound, as do discussions of its association with the marshes, and evidence that it responded to cinchona bark. The pervasiveness of malaria in the marshlands of southeast of England from the end of the sixteenth through the middle of the seventeenth century most probably accounted for the higher crude burial rates recorded there as compared with nonmarshland areas.[26] As today in malaria-infected regions of the world, those who were not struck down by the disease paid a price in terms of reduced capacity to work. In the words of a contemporary observer, George Wither, malaria made prisoners out of those it did not kill.

> In Kent, and all along on Essex side
> A troupe of cruel fevers did reside:
> And around about on every other Coast
> Of severall Country agues lay an Hoast.
> And most of them, who had this place forsooke
> Were either slaine by them, or prisoners tooke.[27]

became major cause of death

Despite evidence that malaria led to an increase in overall mortality during this period, the population of the marsh parishes remained constant or, in some cases, increased. Southeast England was able to attract a steady stream of migrants, people of the "lesser sort"—smugglers, shepherds, or "lookers," as they were known locally—who were prepared to risk their lives in the marshes for the "prospect of gain, and high wages."[28] While it is unclear what drove immigrants to risk their lives in the marshes in exchange for potential economic opportunities, the region may have benefited demographically from the impact of the enclosures occurring elsewhere in England, as the seigniorial class attempted to expand revenues by transforming commons land into pasturage.[29] Historian J. R. Wordie calculated that 25 percent of the enclosures in England occurred during the seventeenth century.[30]

While this immigration helped sustain the population of marsh parishes, it may have fueled the continued transmission of malaria there. Newly infected human hosts, having no prior experience with malaria and thus no resistance to the disease would have been more likely to infect local vectors. As we saw in chapter 1, nonresistant populations experienced higher parasite loads and produced increased levels of gametocytes, which were in turn taken up by feeding anopheline mosquitoes.

[handwritten: 1650 = peak malaria in England]

The Decline of Malaria in England

Malaria probably reached its peak in England around the middle of the seventeenth century, after which it began to gradually disappear. Its disappearance was not the product of any direct effort to control the disease. Rather, it resulted from a continuation of the social and economic processes that led to its emergence during the sixteenth century and was a product of the continued capitalization of agriculture in England. With increasing population and demands for farmland and pastureland, drainage projects increased in number and sophistication over the course of the seventeenth century. Additional capital investments in land reclamation led to the introduction of windmill-driven pumps at the end of the century. With this innovation, drainage projects began to *[handwritten: drains changed everythin]* produce the kind of economic changes that had been promised. It is from that time as well that malaria began to disappear from the southeast of England. The process was slow, and many areas remained conducive to the breeding of *Anopheles* well into the nineteenth century. Not until the introduction of steam-driven pumps in the nineteenth century did lands become sufficiently drained to support widespread increases in cultivation as well as the final elimination of anopheline breeding areas.[31] But by the middle of the eighteenth century, drainage had opened up large areas of coastal land to cattle and sheep farming. This, in turn, contributed to improvements in housing and nutrition. In addition, increases in cattle and their proximity to human populations may have contributed to changes in the feeding habits of local mosquitoes. Mosquitoes may have become more reliant on blood meals from cattle (zoophilic), reducing their biting of humans. With these changes, malaria gradually withdrew.[32]

Shifts in the nature of agricultural production may have contributed as well to the disappearance of malaria from France and the Netherlands from the end of the nineteenth century. In France, new agricultural practices involving the rotation of grain and leguminous crops, along with increases in the practice of keeping cattle in stables, led to the replacement of *A. atroparvus* with other more zoophilic members of the *A. maculipennis* species. General economic improvements, better rural housing, and the migration to urban areas have also been mentioned as possible causes of the disappearance of malaria.[33] In the Netherlands, the progressive reclamation of land from the sea through the building of dikes led to the replacement of *A. atroparvus* by a nonmalaria mosquito, *A. messeae,* which bred more efficiently in the changed environment created by reclamation efforts.[34]

Returning to southeast England, we should recognize that not all the inhabitants of the region benefited from land reclamation and resulting economic improvements. Many of the poor experienced a decline in their well-being, as the reclamations eliminated many of the resources upon which they had depended for their livelihood. In response to these losses, the poor sought employment with those who had benefited from the reclamation, or they turned to the growing textile industry. Others chose to migrate farther afield; going to Ireland or the Americas. A seventeenth-century Fenland proverb described the declining fortunes of Fenmen, many of whom had settled in the Fens in the wake of enclosures elsewhere: "From the Farm to the Fen, from the Fen to Ireland."[35] One of the consequences of this outward movement may have been the transplantation of malaria infections to the emerging American English colonies, where many men and women from the southeast counties traveled in order to find a new life.

The rise and decline of malaria in England reveals again how changing patterns of land use shaped the history of this disease. Efforts to expand agricultural production by altering the flow of water across the land contributed to both the emergence and disappearance of malaria. Yet it is important to recognize that these efforts to increase the productivity of the land were shaped in turn by wider social and economic forces. The expansion of England's population, the growth of cities, the rise of the textile industry, increasing food prices, the capitalization of agriculture, and the movement of human populations all played a role in the changing epidemiology of malaria in England.

MALARIA AND THE NEW WORLD

The same social and economic interests that caused landlords in England to encroach on commons land and attempt to separate the lowlands of southeast England from the sea also drove the quest for new lands on which to expand agricultural production overseas. Poorer farmers pushed off the land, as well as aspiring yeoman planters seeking new lands on which to advance their fortunes, found their way during the course of the seventeenth century to the emerging American colonies of Virginia, Maryland, and South Carolina. Others traveled to New England. And just as the building of embankments created conditions for the transmission of malaria in England, this overseas agricultural expansion contributed to the introduction of malaria to the New World.

It is now generally accepted that malaria parasites came to the New World by ship along with early settlers seeking new lands to exploit. Precisely when and how this occurred, however, is unclear. The accounts of European settlers in the New World before the 1680s do not indicate that malaria-type fevers were a serious problem. This fact has led some to conclude that malaria was brought to the Americas from Africa following the commencement of Atlantic slave trade.[36] Yet, because vivax malaria was endemic in many of the areas from which early European settlers came, they could easily have brought the parasite with them. Many of the early English settlers to the Chesapeake Bay regions of Maryland and Virginia came from London and the eastern counties of Kent and Essex, where, as we have seen, malaria was endemic in the sixteenth and seventeenth centuries.[37] Similarly, French and Spanish settlers along the Gulf Coast may have provided the source for infections in the Deep South and the lower Mississippi Valley, as well as elsewhere in the Caribbean and Latin America.[38] The hibernating ability of *P. vivax*, which permitted it to survive cold temperate climates, also permitted it to cross the Atlantic. The absence of written evidence that early European settlers suffered badly from malaria may simply reflect their knowledge and previous experience with the disease, rather than the absence of malaria infection. The increased severity of malarial fevers recorded in the seventeenth and eighteenth centuries probably followed the arrival of *P. falciparum*, an unfamiliar pathogen for Europeans because many of the common species of anopheline mosquitoes in western Europe were resistant to falciparum infection, including the main transmitter of malaria in England, *A. atroparvus*.[39] In

all likelihood, Africans brought to the New World as slaves introduced *P. falciparum* into the Americas.

In the New World, malaria parasites met a ready population of
anopheline mosquitoes, capable of being infected by and transmitting both *P. vivax* and *P. falciparum* parasites. Among these *Anopheles quadrimaculatus,* native to low country areas of the southern
Atlantic seaboard, was especially susceptible to malaria parasites.

While the coming together of Old World parasites and New
World mosquitoes made possible the introduction of malaria into
the New World, its subsequent expansion was dependent upon
the creation of ecological conditions conducive to malaria transmission. In other words, the geographical origins of parasites does
not explain why the disease took hold in particular places and at
particular times. To answer this question we must look here—as
we did in Sardinia, Rome, and England—at the ecological transformations that human settlements produced.

But first we must stop for a moment and address the question
of malaria's impact on the native-born populations of the New
World. If malaria arrived in the Americas for the first time along
with early European settlers, it may very well have taken a heavy
toll among the immunologically naive native populations living in
lowland areas of the Americas, joining other Old World diseases
(smallpox, influenza, measles, and tuberculosis) in decimating the
Amerindian population. Unfortunately, the data available do not
allow us to make fine distinctions among the various scourges
that accompanied European colonization. The combined impact
of imported pathogens on Amerindian populations is a source of
considerable debate, and we can only speculate about the added
burden contributed by malaria.

Identifying the precise set of circumstances that gave rise to
malaria over broad areas of the Americas is impossible. We know,
however, that the development of rice production in the tidewater areas of the Carolinas, tobacco in Maryland and Virginia, and
sugar in the Caribbean and Louisiana all contributed to the early
emergence of malaria. A careful examination of the rise of malaria
in the emerging plantation economy of South Carolina illustrates
the complex ways in which malaria was associated with the expansion of cash-crop production in the Americas.

Rice, Slaves, and Malaria in Colonial South Carolina

English settlers first established themselves in what would become the colony of South Carolina in 1670. During the first 10 years settlers expanded their holdings, creating plantations adjacent to the Ashley and Cooper Rivers and in the low-lying coastal swamps. Reports from this period give little indication that diseases of any sort were a major concern to the new inhabitants. To the contrary, the emerging colony was described as being extremely healthful. A report from 1682 noted that in July and August the inhabitants "have sometimes touches of Agues and Fevers, but not violent, of short continuance, and never Fatal. English children there born are commonly strong and lusty, of Sound Constitutions and fresh ruddy complexions."[40]

This was certainly a far cry from the conditions in the coastal counties of southeast England, from which many of these settlers came. While some of the positive views of the region's healthful climate can be attributed to propaganda produced by backers of colonization to encourage settlers to move to the colony, there is reason to believe that the colony was reasonably healthy. Little in the historical record indicates the presence of malaria to any degree.

Research on the distribution of anopheline mosquitoes in South Carolina during the early decades of the twentieth century suggests a possible reason for this apparent lack of malaria during the early years of settlement. The areas in which settlers first established plantations did not support the breeding of the *A. quadrimaculatus* mosquito, which was the dominant vector for malaria throughout the eastern United States. Early settlers engaged in mixed farming and cattle raising along the Ashley and Cooper Rivers and next to the coastal swamps. These activities disrupted the land to some degree, but they did not produce large collections of standing waters containing rotting vegetation that *A. quadrimaculatus* preferred.[41] Moreover, because *A. quadrimaculatus* avoided salty or brackish water, it would not have been present in the coastal marshes. It is also unlikely that *A. quadrimaculatus*, with a short flight range, would have traveled in any great numbers from inland swamps to the coastal river settlements.[42]

Malaria, however, must have occurred to some degree among early settlers for the parasites to survive during the first decades of settlement. Perhaps a second local vector, *A. crucians,* transmitted malaria among the settlers.[43] Unlike *A. quadrimaculatus, A. crucians* breeds readily in the salty coastal marshes, but it is less susceptible to infection and thus a less efficient transmitter of malaria. This would explain how malaria could have persisted without becoming a significant health problem.[44]

Whatever malaria occurred in South Carolina before the end of the seventeenth century was probably *P. vivax,* and thus unlikely to have produced many deaths. The early settlers, moreover, would have been familiar with the disease and may not have taken much note of it. There is no evidence of the more severe forms of fever associated with *P. falciparum* malaria, even though African slaves began to be imported into the colony as early as 1670. These slaves were imported from Bermuda and Barbados, islands that were malaria free.[45] *swampy/ lots of ponds = great for malaria*

If the lack of conditions suitable for the production of efficient malaria vectors accounts for the relative absence of malaria fevers among early settlers, changes in these conditions help explain the rise in malaria that occurred at the end of the seventeenth century. In the 1680s settlers began to push inland toward the headwaters of the Ashley and Cooper Rivers. During this same period rice growing began to take hold. The bulk of rice production occurred in inland swamps, where rice planters depended on ponds or reservoirs of fresh water drawn from swamps or rivers to irrigate their crops. The construction of these ponds and reservoirs provided ideal breeding places for *A. quadrimaculatus* mosquitoes and thus laid the basis for the expansion of vector populations and the transmission of malaria.[46] To further the problem, inland rice planters developed the habit of building their houses in close proximity to their fields, reportedly to keep an eye on their workers.[47] Thus inland rice production increased vector breeding and exposed human populations to malarial infections.

Inland swamp cultivation was also labor intensive. Before rice could be planted, swamps had to be drained, irrigation ditches dug, and embankments built to protect fields from flooding. Once planted, rice cultivation required endless days of weeding, as the irrigation water that nourished the grain also encouraged weeds to grow. Finally, the rice had to be harvested. The increased demands for labor led settlers to seek additional slaves to supplement the original numbers that had been brought from Bermuda and the Caribbean.[48] While the numbers of African slaves were small at first and were supplemented with indentured European servants, the African population increased steadily through the early decades of the eighteenth century. By 1708 there were more Africans in the colony than Europeans. Most of the slaves continued to come from the Caribbean. Only 15 years later, however, trade with Africa began in earnest, and the number of slaves imported directly from Africa increased rapidly over the next three decades. Roughly 30,000 slaves were imported into South Carolina from Africa between 1720 and 1740.[49] The increased importation of slaves directly from Africa during this period probably led to the domestication of *P. falciparum* malaria in South Carolina, as many of the Africans brought to the New World came from areas where *P. falciparum* was endemic. While these enslaved Africans probably possessed some form of acquired or hereditary resistance to *P. falciparum,* they could still become infected with the parasite and transmit it to others. However, human hosts who exhibit resistance to *P. falciparum* are less efficient transmitters of the parasite to anopheline mosquitoes than persons without resistance. This being the case, white settlers were probably more responsible for the subsequent transmission of *P. falciparum* malaria in South Carolina than were enslaved Africans.

Malaria took a dreadful toll among settlers during the early and middle decades of the eighteenth century. Infant and child mortality was high in all the low-country parishes. In Christ Church Parish, 86 percent of all those whose births and deaths are recorded in the parish register died before the age of 20; 33 percent of the females and 57 percent of the males died before the age of 5. A large proportion of these deaths occurred between the months of August and November, when "fevers and agues" were most intense. Of all the children dying before their first birthday, 90 percent died during these months, as did 77 percent of all persons who died before reaching the age of 20.[50] Taken together with descriptions of the fevers and chills experienced by residents during these months, these numbers strongly point to malaria as the major cause of death in the low country (although yellow fever outbreaks certainly contributed to this carnage). Because mortality statistics for persons of African descent were not systematically recorded, it is difficult to compare their death rates with those of persons of European descent, although it was generally believed at the time that Africans suffered less than Europeans did from yellow fever and malaria and that this advantage made Africans particularly useful for plantation labor.[51]

While malaria remained a health problem in South Carolina well into the twentieth century, there is evidence that its intensity diminished during the last quarter of the eighteenth century. Life expectancy rates improved significantly over those of earlier decades for both men and women during this period. Cultural adaptations among the inhabitants contributed in part to this improvement. Planters came to see the swampy inland environment as responsible for their fevers and sought shelter along the coast or on higher ground further inland during the fever seasons. The wealthiest planters spent the fever season in northern resort towns such as Newport, Rhode Island.[52] Their white overseers and slaves were not so privileged. ⤳why did malaria fizzle out

This declining mortality from malaria could not be attributed to medical advances, although cinchona bark (known locally as Jesuit bark or Peruvian bark) also became an effective treatment for fevers and agues, as it had in England. However, its expense limited its use. It was not until quinine, its active ingredient, was isolated and made widely available that South Carolina residents had an effective medical treatment for malaria. But this did not happen until the early nineteenth century, whereas the drop in malaria deaths began toward the end of the eighteenth century.

Changes in the methods of rice production may have been a more important contributor to declining malaria mortality. The difficulties of both irrigating and weeding inland swamp rice fields led some planters to seek an alternative system of production. As early as the 1730s they noted that the diurnal rising and falling of coastal rivers, caused by ocean tides, provided a new way of irrigating rice fields. By situating one's fields along riverbanks protected by embankments, and then constructing flood gates that would

allow the rising tide to gently flood the fields and the lowering tide
to remove the water at regular intervals, a planter could solve the
problem of both irrigation and weeding. Regular flooding provided fields with water, while preventing the growth of weeds.

Tidal irrigation did not become widespread until after the
American Revolution, because construction of the new irrigation
system required considerable capital. Rising demand and prices
for grain in Europe following the Revolution made it possible for
a growing number of planters to shift to tidal rice cultivation.[53]
Still, not all planters could afford the investment needed to make
the transition. Many remained wedded to swamp cultivation.
Those who made the transition increased their profits. As a result,
plantation society became increasingly stratified on the basis of
wealth as tidal rice production expanded.

The shift to tidal cultivation had two positive consequences for
the problem of malaria. First, those planters who adopted the new
technology no longer lived in the inland swamp areas that were
particularly suited to the breeding of *A. quadrimaculatus* mosquitoes. Second, routine flooding of tidal rice fields disrupted the
breeding of local vectors. Together, these changes reduced transmission of malaria on the tidal cultivation plantations. Obviously
those who continued inland swamp cultivation, a group of increasingly poorer planters, would have continued to be susceptible
to malaria. Unfortunately, we do not have the comparative mortality data needed to judge whether malaria mortality was higher
on the interior plantations than the tidal plantations. Even if we
did, it would be difficult to differentiate the effects of changed
methods of cultivation from those caused by differences in wealth.
One would assume that the wealthier coastal planters would have
been more likely to sojourn away from the low country during
the fever months and that they would have had greater access
to cinchona bark. Both factors would have reduced the impact
of malaria independently of the change in cultivation methods.
Still, it seems likely that the capital investments in new forms of
cultivation played a role in the reduction in malaria mortality that
occurred in the last quarter of the eighteenth century, just as it had
in England earlier in the century.

Nevertheless, malaria remained a problem for a large portion of the population living in areas further up the rivers and in the marshy hinterlands just beyond the coast. And by the end of the eighteenth century malaria had spread further inland, becoming a seasonal problem for residents living in the piedmont areas of the colony. In addition, the Civil War dramatically disrupted rice production on the coastal plantations and was accompanied by a resurgence of malaria there and elsewhere in the South.

Malaria Moves West   ⤷ civil war also had influence

The end of the American Revolution saw the opening up of the vast lands lying west of the Atlantic seaboard and an acceleration of the westward movement of colonial settlers. This post-Revolutionary movement also contributed to the westward spread of malaria. Settlements along the upper Mississippi Valley appear to have been relatively healthful and fever free before the Revolutionary War. As new settlers arrived from the East during the last quarter of the eighteenth and early nineteenth centuries, however, they brought with them malaria parasites. The new settlers also created social and environmental conditions that facilitated the transmission of malaria.   followed river

Settlers who moved west followed river courses inland, establishing settlements beside the rivers in order to ensure a supply of water and access to river transport. The soils of the river bottomlands were also more fertile than those further inland. Yet the bottomlands also provided breeding sites for *A. quadrimaculatus,* which was either indigenous to these areas or expanded westward, taking advantage of the ecological conditions created by the settlers. Once settlers established their beachhead along the rivers, they began clearing the forests to make room for farmlands. In doing so, they exposed pools and streams to sunlight, creating additional breeding grounds. They also dammed up streams to create reservoirs in which *A. quadrimaculatus* preferred to breed. The poor quality of the housing that marked most early settlements also facilitated the spread of malaria. Without glass windows or proper screening, malaria vectors had easy access to the inhabitants. Farm animals were also in short supply and, when present, were often let loose to forage on their own rather than housed in shelters near the farmers' houses. There were thus few alternative

lifestyle encouraged spread of malaria

hosts available close by to deflect the mosquitoes away from the settlers. As a result of these various ecological factors, the residents of the early river towns and farms were subject to recurrent malaria attacks in the summer and autumn months.[54] In this way malaria spread westward along the Ohio, Tennessee, and Missouri Rivers, and up and down the Mississippi, from New Orleans to Minnesota and Canada.

Reports of malaria fevers can be found in abundance in these areas from the 1760s onward. According to Boyd, who examined reports of the disease throughout the region, "By the 1850s, malaria was extensively endemic throughout the United States with hyperendemic areas in the southeastern states, the Ohio River Valley, Illinois River Valley and practically all of the Mississippi River Valley from St. Louis to the Gulf."[55] Known variously as Arkansas chills, swamp fever, the ague, or simply "the shakes," malaria was a reoccurring fact of frontier life. On muggy August nights, those afflicted with the chills were said to huddle around roaring fires.

By the 1860s malaria began to disappear from various parts of the upper Mississippi Valley as well as from the northeastern and Atlantic states, though not without periodic resurgences. Just as in England, a series of social and economic transformations combined to reduce human exposure to malarial parasites, while at the same time improving their ability to resist infections. These changes, associated with the expansion of commercial agricultural production, reversed the conditions that had contributed to the initial expansion of malaria. A brief look at the history of malaria in Illinois reveals how changing patterns of settlement and production contributed to the rise and decline of malaria in the American Midwest.

Trans-Appalachian migrants from the Piedmont and Great Valley, ranging from Pennsylvania through Virginia to the Carolinas, explored and settled the American Bottom, near St. Louis, and the Shawnee Hills of southern Illinois at the end of the eighteenth century.[56] By the 1830s they had established settlements northward along both water and land routes. Somewhat later, settlers from the Middle Atlantic and northeastern states settled the northern parts of Illinois.

During the initial colonizing process, isolated farmsteads were

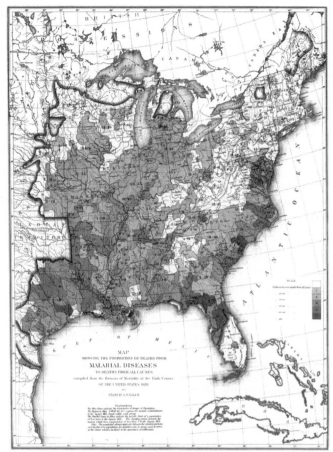

not very prevalent in North

Malaria Map of the Eastern United States, 1870s
Source: U. S. Department of Interior, Census Office, *Statistical Atlas of the United States, Based on the Results of the 9th Census, 1870* (New York: Julius Bien Lith., 1874), pl. XLII.

scattered in remote woodland frontiers. Homesteads based on loose family and clan settlement clusters were dispersed less than a mile apart. Families lived by hunting with some patch agriculture. Lack of transport limited efforts to produce surplus crops for market, so that most production was subsistence based. Nonetheless, by the 1830s a growing number of yeoman farmers were transporting their crops by flatboat or keelboat to sell in emerging river towns, particularly in St. Louis.

summer = main time [handwritten margin note]

Malaria was intense during the summer months within almost all of the settlements in southern Illinois, but particularly so along river courses or where water had been impounded.[57] Even away from river courses in eastern and central Illinois, poor drainage made malaria a problem as well. Epidemiologist V. C. Vaughan claimed in 1923 that in the 1860s the rural population of southern Illinois was almost universally infected. Describing the summer of 1865, he wrote that "every man woman and child in southern Illinois, at least within my range shook with the ague every other day."[58]

Between the 1830s and 1850s, midwestern agriculture experienced vigorous expansion and reorganization. Corn production linked to the raising of cattle and hogs spread from western Virginia, through southern Ohio, into southern Illinois.[59] The growing commercialization of agricultural production in Illinois was supported by the expansion of industrial and agricultural marketing centers in Chicago and St. Louis, as well as by increased demand for surplus grain and meat production in the industrial northeast corridor from Boston to Baltimore. Improvements in transportation linked Illinois farmers to these industrial markets by steamboat, roads, canals, and an expanding railway system. By 1850 most of the farmers living in the western half of Illinois, along the Mississippi and Illinois Rivers, were engaged in commercial farming.[60]

The commercialization of Illinois agriculture was achieved through an intensification of production using family labor and increased capital investments in labor-saving farm machinery. It also occurred without producing a surplus of poor and landless people. Those who did not make a living off the land did so in the towns that were scattered through the region. Very few rural families supplied workers for the growing industrial center in Chicago, which relied largely on foreign immigrants for its workers.

A number of social and ecological changes accompanied this agricultural transformation, and these changes contributed to the gradual reduction of malaria during the second half of the nineteenth century. Improvements in transportation and market demand led an increasing number of farmers to expand their cultivation away from the river bottoms into the prairie hinterlands, thereby reducing their exposure to malaria-carrying mosquitoes. Commercial farming also led to improvements in housing, which further reduced exposure to malaria. John Peck in his guide to settlement in the West noted how the second wave of commercial farmers in the Midwest constructed sturdier log homes with glass windows and brick or stone fireplaces.[61] In eastern Illinois, the draining of farmlands increased agricultural production and reduced malaria transmission. Finally, the widespread combination of corn and livestock production, increased the population of livestock associated with human settlement and thus provided alternative vertebrate hosts for local anopheline mosquitoes and particularly for the dominant transmitter of malaria, *A. quadrimaculatus*. With the increased availability and use of quinine, malaria became even less of a problem at the end of the nineteenth century. While cases of malaria still occurred, the impact of the disease was greatly diminished up and down the upper Mississippi Valley.

❖ ❖ ❖

Malaria reached its furthest northern expansion during the nineteenth century, with outbreaks in northern Russia and Canada. Yet 100 years earlier it had already begun to disappear from parts of northern Europe, as we have seen in the cases of England, France, and the Netherlands. It had also begun to retreat from many parts of North America after 1850. This withdrawal occurred in the absence of any direct efforts by human populations to curtail the disease. Rather, changing social and economic conditions, associated with capital improvements in agriculture, transformed the ecological relationship among malaria parasites, human hosts, and anopheline mosquitoes, resulting in a decline in the burden of malaria. These transformations had allowed local populations, in a sense, "to grow out of malaria."

The retreat was not uniform, as some northern areas continued to have outbreaks of malaria into the early twentieth century. For example, malaria increased in intensity in the Central Valley of California during the early twentieth century. The expansion was driven by the widespread use of irrigation, together with the rapid construction of roads and railways, which created microenvironments conducive to mosquito breeding.[62]

Nor did malaria's retreat from the North occur without set-backs. In New England and the Middle Atlantic states, malaria was disappearing during the middle decades of the nineteenth century but increased toward the end of the century. This surge was in part due to returning Union soldiers who had been infected while serving in the South during the Civil War. But other changes, including the building of dams and reservoirs, may have also contributed to this increase. Irish and Italian immigrant laborers constructing the Cochituate and Wachusett Reservoir aqueducts to Boston were ravaged by malaria in 1896.[63] Nonetheless, by the time medical scientists unraveled the etiology of malaria, allowing public health officials to develop strategies for eliminating the disease, malaria was well on its way to becoming once again a tropical disease.

CHAPTER THREE

A Southern Disease

While malaria gradually receded from the North, it continued to maintain its grip on populations living in tropical and subtropical regions of Africa, Asia, and Latin America during the nineteenth and early twentieth centuries. Malaria also remained entrenched in the milder temperate areas of the American South and southern Europe. In this and the following chapter, we examine the causes of this persistence.

Ask a malariologist today why social and economic transformations contributed to a decline in malaria in the temperate North but not in the warmer regions of the South and you will likely be told that the answer lies in differences in climate, the nature of malaria transmission, and the efficiency of local malaria mosquitoes. Warmer, moister climates facilitated the breeding of anopheline mosquitoes, as well as the reproduction of malarial parasites. Malaria in much of the tropics was stable, with transmission occurring over most of the year. By contrast, malaria transmission in the temperate areas of the globe was more markedly seasonal with long periods during which colder temperatures reduced vector populations and interrupted transmission. The reduction in vector populations made it easier to eliminate vectors and interrupt transmission. Finally, the efficiency of a number of tropical mosquitoes in transmitting malaria, due to their susceptibility to infection and their strong preference for the blood of human hosts, meant that very few mosquitoes needed to survive in order to maintain transmission.

All of this is certainly true. Yet the answer leaves out another critical difference. Patterns of agricultural development in much of the so-called tropical world were fundamentally different from those that occurred in northern Europe and upper regions of the United States from the end of the eighteenth century. These differences had important consequences for the human ecology of malaria. In the South, agricultural development both failed to provide local populations with the opportunity to "grow out of malaria" and all too often exacerbated conditions of exposure, leading to an expansion of malaria transmission.

The economic history of tropical and subtropical regions of Africa, the Americas, and Asia from the end of the eighteenth century through the middle of the twentieth century, like that of Europe and North America, was marked by extensive efforts to expand agricultural production. Agriculture also expanded in the American South and southern Italy during the second half of the nineteenth century. However, agricultural expansion in the South and the North differed in how land, labor, and capital were mobilized and in how agricultural markets were structured. These differences had ecological consequences, which, in combination with disparities in climate, vector populations, and transmission patterns, sustained, or in some cases increased, human exposure to malaria in the South. This chapter examines why malaria persisted in the American South and Italy at the same time that it was disappearing in much of the rest of western Europe and North America. Why were these two areas different? In chapter 4, we examine how patterns of tropical agriculture encouraged malaria transmission within colonized regions of Africa, Asia, and Latin America.

MALARIA AND RECONSTRUCTION IN THE SOUTHERN UNITED STATES

Malaria devastated combatants and civilian populations living in the South during the American Civil War. The movement of immunologically naive northern troops into malaria-endemic areas of the South, combined with unsanitary camp conditions that encouraged the breeding of anopheline mosquitoes, created ideal conditions for the transmission of the disease. Union troops suffered multiple episodes of malaria, producing an extraordinary morbidity rate of 2,698 cases per 1,000 troops between May 1861 and June 1866. The total number of recorded cases was 1.3 million, with more than 10,000 deaths.[1] Although the exact number of Confederate army deaths from malaria is not known, there were 41,539 cases in an 18-month period (January 1862 to July 1863) in South Carolina, Georgia, and Florida.

The war also created conditions that fostered the spread of malaria among civilian populations. Agricultural production over wide areas of the South was seriously disrupted by the war. Lands that had been cultivated were left untended, allowing for the collection of standing water and the increased breeding of malaria vectors. When soldiers returned after the war to restart their planting, they often found a changed landscape in which malaria-carrying mosquitoes bred in large numbers. Dr. Henry R. Carter, an eminent American sanitarian who devoted much of his professional life to combating yellow fever and malaria in the United States, described conditions on his family's Virginia plantation at the end of the Civil War. Carter, who was 13 when the war ended, remembered that "There was a great increase in malaria after the Civil War in my section of Virginia—just above tidewater. This country had never been free of malaria; but on our plantation at least, it had not been of sanitary importance for a number of years. After the war several of the family had it—I and two little sisters—and it was fairly general, though not severe, all over our neighborhood."[2]

Carter ascribed the upsurge in malaria to the wartime disruption of cultivation and the clogging of drainage ditches. Of additional importance were poor diets, the confiscation by Union troops of farm animals that had served as alternative blood sources for local mosquitoes, and the Union blockade of southern ports that had reduced access to quinine.

Despite these setbacks, southern farmers attempted to rebuild their lives after the war, restarting farms and expanding cultivation. As a result, the last decades of the nineteenth century saw an expansion of agricultural production over broad areas of the

South. Much of the land that was put into production was devoted to cotton. Yet patterns of production in the postwar South differed from those that occurred during the same period in the North. Northern agriculture from the 1880s was increasingly capital intensive, mechanized, and based on the labor inputs of relatively small numbers of agricultural workers, most of whom owned their own land. By contrast, over wide areas of the American South agricultural production was marked by the extensive use of land, limited capital investments, and a heavy reliance on cheap labor rather than machinery. Where northern agriculture contributed to a decline in malaria, southern patterns of production sustained and frequently increased conditions of exposure, leading to an expansion of malaria. Like all generalizations, this one is subject to local exceptions. As we will see, certain forms of agricultural production in the South reduced malaria. Nonetheless, the cotton economy that dominated much of the South following the Civil War contributed to the persistence of malaria.

Cotton had been grown on plantations and small farms in the lower South from the Carolinas through southern Georgia, Alabama, and Mississippi before the Civil War. Production declined during the war as white landowners abandoned their farms to fight in the war. Moreover, as noted earlier, many farms were also damaged by the conflict. A total of 4.5 million bales were grown in 1861, 1.5 million in 1862, 500,000 in 1863, and only 300,000 in 1864. As a result, cotton prices rose and remained high during the immediate postwar period. Understandably, many farmers, desperate for capital to resurrect their farms, saw cotton as offering the quickest return on investment and began expanding the number of acres devoted to its production. When prices declined in the 1870s and 1880s, these same farmers chose to expand production even further, hoping increased production would compensate for declining prices, allowing them to maintain their income and standard of living. By the 1880s cotton had replaced most other crops, creating a monoculture economy over wide areas of the South. Cotton farming also expanded westward into new areas of production in Louisiana, Arkansas, and Texas during this period.[3]

The lack of capital that led southern farmers to expand their cotton production also prevented them from investing in labor-saving technologies and other means of increasing productivity. The expansion of production therefore depended on the use of large numbers of workers. Smaller farms, dependent on family labor, found it difficult to expand their labor supply and thus production. On larger plantations, the war and the emancipation of slaves disrupted the supply of labor. No longer legally bound to their former owners, many slaves moved away from the farms on which they had worked before the war, believing that this was the only way to ensure their freedom. Some sought to acquire their own farms, fulfilling the promise of "forty acres and a mule" made by Reconstruction politicians. Only about 20 percent succeeded in achieving this goal; but this 20 percent represented a significant reduction in the available work force. Some former slaves moved out of the South altogether, though out-migration was probably less than 2 percent per decade in the 1870s and 1880s. Others moved westward in hopes of gaining better terms of employment in the new cotton-growing areas of Arkansas, Mississippi, and Texas, creating labor shortages on the farming areas they left behind.

Faced with an uncertain labor supply, planters sought new ways to reestablish their access to black workers without having to pay high wages. To do this they negotiated various kinds of tenancy arrangements with their former slaves and other freedmen.[4] These arrangements ranged from sharecropping, a kind of rent tenancy in which the farmer provided tenants with access to land in exchange for a portion of the tenant's harvest, to labor tenancy, where the tenant worked on the farmer's land in return for a piece of land for growing his own crops. Many blacks hoped to use these arrangements to accumulate capital and purchase land of their own. Yet few did. The terms of most contracts made it difficult for tenants to accumulate capital. Sharecropping contracts typically left the tenants with less than half of the crop they produced. Labor tenancy contracts placed high demands on the labor of tenants and their families, leaving them little time for growing their own crops. Planter-dominated legislatures passed laws in some states that restricted the ability of black tenants to break tenancy contracts and seek better terms from another planter.[5] They also made it illegal for planters to compete for the services of black farmers. Both sets of laws restricted the labor market and the ability of tenants to improve the terms of their contracts.[6]

Blacks were also trapped by indebtedness. Tenant farmers frequently needed to purchase tools and seeds and other supplies from local merchants or planters, who would advance them credit against the tenant's share of the harvest. A combination of high prices charged for these goods plus interest charged on outstanding debts prevented many tenant farmers from repaying them in a single season. Because tenants needed new supplies the following season, blacks found themselves continually in debt. The continued burden of debt, combined with laws that made failure to repay a debt a crime, kept tenants in a state of perpetual debt peonage for long periods of time, while ensuring their continued poverty.[7]

Blacks were not the only group to become enmeshed in this system of tenancy and debt peonage. Poorer whites, who like their wealthier neighbors devoted their lands to cotton growing, found that in bust years they were unable to make ends meet, forcing them to borrow money from merchants and wealthier planters. During the 1880s, when cotton prices bottomed out, many poor white farmers were forced to sell their land and join the ranks of former slaves as tenants or sharecroppers.

Even wealthier planters paid a price for their dependency on labor tenancy. Labor tenants, lacking ownership of the land, had little vested interest in improving it. Thus conditions of production were often poor. Drainage was not maintained and fences were not repaired. These conditions reduced the productive capacity of many plantations. They also contributed to the spread of malaria.

The system of agricultural production that emerged around the cultivation of cotton in the rural South following the Civil War encouraged the transmission of malaria in a number of ways. To begin with, most sharecroppers and labor tenants were dirt poor and remained that way through the 1930s. Continually in debt, they lacked the resources needed to make improvements in their houses or to maintain proper sanitation. Their houses were poorly constructed and lacked adequate protection from invading mosquitoes, which bred in neighboring ponds and other collections of water that formed in the rich bottomlands in which cotton grew best. Overcrowding was the norm, with as many as 10 to 12 people occupying one small house. In these conditions it did not take long for a single infected child or adult to become the source for the transmission of malaria throughout a household. Diets were heavy in fats and carbohydrates and low in minerals, vitamins, and protein. Nutritional deficiency diseases such as pellagra were widespread in the South. Poor diets and other parasitic infections, including most commonly hookworm, produced anemia and reduced overall energy and health.[8] When malaria parasites attacked the red blood cells of already anemic individuals, the health consequences could be severe. Once ill, few sharecroppers could afford to purchase medicines or consult a physician. Because quinine in particular was often out of the reach of most sharecropper families, they relied on home remedies or just accepted "autumnal" chills and fevers as part of life. As in many parts of the tropical world today, malaria had its greatest impact on infants and young children.

While the conditions under which sharecroppers lived were generally poor, they became worse when cotton prices bottomed out. As the price of cotton fell, sharecroppers' ability to maintain their homes, feed their families, and acquire medicines when sick declined. There is some evidence that malaria mortality in the Mississippi Delta fluctuated in line with rising and falling cotton prices. The Mississippi Delta possessed rich soils and attracted farmers seeking to take advantage of expanding cotton markets. Yet the area possessed many low-lying swampy areas that provided ideal breeding sites for *A. quadrimaculatus* mosquitoes. The risk of malaria on the Delta plantations that began growing cotton in the 1870s was so high that landowners seldom lived on their plantations, hiring white overseers to run them in their stead. Poorer white farmers also avoided the region, and the production of cotton was done almost exclusively by black tenant farmers who migrated to the Delta in hopes of obtaining better terms of employment.[9] Between 1916 and 1926 cotton prices in the region fluctuated between a low of 18¢ and a high of 32¢ per pound. Malaria mortality, while generally declining, fluctuated as well between a high of 1,426 deaths and a low of 365 deaths during these

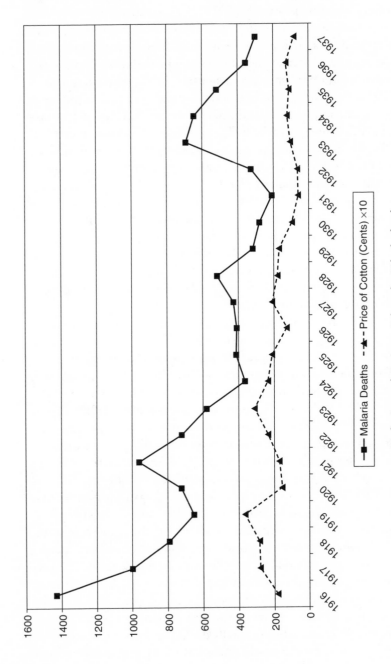

Malaria and Cotton Prices in Mississippi

Source of data: W. B. Brierly, "Malaria and Socio-Economic Conditions in Mississippi," *Social Forces* 4 (1945): 451–59.

same 10 years. There was a strong negative correlation between the price of cotton and the number of malaria deaths over this period. Declines in the price of cotton were followed by increases in malaria mortality, whereas rising cotton prices were associated with declining mortality.[10]

Malaria was not unique to cotton-growing areas but could also be found in connection with other crops. Thus, Dr. Henry Carter noted that while conditions in the Virginia Piedmont improved between 1865 and 1870, and malaria retreated to the low-lying tidewater areas to the east, the disease returned in association with the rise of tobacco production during the 1870s.[11]

Before the Civil War, wheat and corn production dominated the economy of the piedmont region in which the Carter family plantation was located. Wheat was produced for market and corn to feed mules, slaves, and horses, which were used to plow the land. During the 1870s, however, wheat production expanded into Iowa, Minnesota, and the Dakotas. The construction of land-grant roads and the reduction in transport rates allowed midwestern wheat to compete in eastern grain markets. Wheat prices dropped accordingly, and the Piedmont economy experienced several years of depression. In response to declining profits, Virginia farmers moved out of wheat production and into the growing of tobacco and cattle raising. Corn production also decreased as tobacco farmers required fewer mules and horses for plowing.

These changes in production had ecological consequences, which contributed to the resurgence of malaria. Unlike wheat, tobacco was grown on small parcels of land. This meant that large areas on which wheat and corn had previously been grown were left fallow. While wheat had been produced on high well-drained lands, corn was produced in bottomlands, where water could accumulate if fields were not properly drained. As farmers abandoned corn production, these bottomlands were allowed to become waterlogged following heavy rains, creating breeding sites for *A. quadrimaculatus* mosquitoes. Increases in cattle might have deflected mosquitoes from humans; however, the cattle were not housed in barns but left to graze in pastures located at some distance from human habitations. Malaria increased and, according to Carter, remained a problem up through the 1920s.

Not all areas of the South suffered from malaria during the last decades of the nineteenth century. Rice plantations in Arkansas and Texas remained relatively free of malaria, despite their use of irrigation and the presence of an abundance of malaria vectors.[12] Writing in 1926, M. A. Barber marveled at the fact that rice-growing areas were plagued by conditions that produced immense numbers of anopheline mosquitoes and yet there was little malaria.[13] The absence of malaria was particularly surprising, given the tendency for malaria to accompany rice growing in other parts of the globe, including South Carolina, Spain, and northern Italy.

The difference between rice growing in the prairie rice regions and rice production in South Carolina, as well cotton production in the rest of the South, was the means by which rice was grown in these southwestern states.[14] Rice production was highly capitalized in Arkansas, Louisiana, and Texas. It was developed by land speculators and farmers from the Midwest who moved into the region and adopted production methods and technology used for growing midwestern wheat. The new rice farms were not the one- or two-mule sharecropping operations characteristic of farming in most of the South. With larger acreages, the rice growers in Louisiana, Texas, and Arkansas adopted gangplows, seeders, discs, the twine binders for harvesting, and steam engines to power their threshing machines. Irrigation methods were perfected and farmers introduced better breeds of rice and improved milling technology in the 1890s. By the early twentieth century, rice growers in these three states were the most modernized farm operators in the South.[15]

The heavy capitalization and mechanization of rice growing eliminated the reliance on large numbers of poorly paid agricultural workers. No large rural population lived near the fields, exposed to anopheline mosquitoes. Rice growers relied on a small number of workers, who were relatively well paid and lived in towns in above-average houses, many of which had screens. Workers who contracted malaria, moreover, had access to quinine. Barber concluded that the social and hygienic conditions in the rice regions were generally superior to those found in the rural regions of other parts of the southern states. He noted that malaria rates were generally higher in regions outside the rice-growing areas.

In short, rice production in the South resembled wheat and corn production in the Midwest much more than cotton production in the rest of the South. The exceptionalism of malaria and rice production supports the argument that it was the South's sharecropping, cotton-dependent economy, as much as its warmer climate, that was largely responsible for malaria's persistence following the Civil War.

Malaria remained a problem throughout the South until the 1930s. Efforts at malaria control played a role in restricting the disease in the 1910s and 1920s. However, these efforts were centered primarily on towns and cities. Little was done to eliminate the disease from the surrounding countryside. Nonetheless, malaria began receding along a broad front during the 1930s. This change was due in large measure to the elimination of the conditions of agricultural production that had dominated much of the South since the end of the Civil War.

Ironically, the boll weevil, an insect that destroyed cotton plants and became the bane of southern cotton growers, helped eliminate malaria. The march of the boll weevil across the South, beginning in Texas in the 1890s, contributed to the demise of the cotton industry. Yield losses associated with the boll weevil in Georgia reduced cotton acreage by 50 percent from a historic high of 5.2 million acres during 1914 to 2.6 million acres in 1923. Production dropped from 2.8 million bales to 600,000 bales over the same period.[16] Many sharecroppers were forced to abandon agriculture during this period. Some moved to the West, others to the North. Nearly a million and a half blacks left the South between the 1890s and the 1930s.[17] As they moved out of farming, flocking to towns and cities, or emigrating out of the South, they withdrew from the conditions of exposure that had fueled the transmission of malaria. In short, the boll weevil separated former sharecroppers from the anopheline breeding grounds. As malaria control measures took hold in towns and cities, moreover, these urban sites became refuges from rural malaria.

The final blow to the old forms of cotton production was the passage of the Agricultural Adjustment Act by the U.S. Congress in 1933. As Margaret Humphreys has shown, the act paid farmers to take land out of cotton, tobacco, and rice production and provided much-needed capital for the purchase of machinery. Cotton production, as well as the farming of new crops, such as peanuts, became increasingly mechanized. Landowners, no longer reliant on sharecroppers for production, canceled tenant contracts, forcing most of the remaining sharecroppers off the land. Many of them remained dependent on agricultural employment, but they no longer lived on the land they worked.[18] Instead, they joined those driven off the land by the boll weevil and resided in towns and cities, returning to the fields to work as day laborers. Thus, the decline of malaria in the South was tied to conditions that undermined the production of cotton and resulted in the capitalization of agriculture. From the end of the 1930s, forms of agricultural production in the South came to resemble more and more those that had developed in the American Midwest at the end of the nineteenth century. With these changes, the human ecology of malaria changed.

LAND, LABOR, AND MALARIA IN SOUTHERN ITALY

At roughly the same latitude, thousands of miles east, conditions of agricultural production also allowed malaria to maintain its hold in southern Italy. By the end of the nineteenth century, malaria had disappeared from much of western Europe. Pockets of the disease remained in various parts of the Netherlands, Greece, and Spain, and outbreaks occurred occasionally in England. Yet for the most part, ecological transformations linked to growing capital investment in agriculture combined to push back malaria. By contrast, malaria remained a major health problem over much of the Italian peninsula.

In the North of Italy, malaria persisted in conjunction with the production of rice in Piedmont and Lombardy, as well as in the Veneto and Emilio-Romagna. Unlike rice production in Arkansas, Texas, and Louisiana, rice growing in northern Italy was labor intensive, involving little in the way of mechanization. The worst malaria, moreover, occurred in the fields in which production was the least capitalized. Wealthier farmers employed methods of irrigation that insured a constant flow of water that reduced the breeding of anopheline mosquitoes, whereas poorer farmers simply impounded water for extended periods of time, creating stagnant basins in which larvae grew in large numbers. Everywhere, rice production depended on migrant workers who were recruited from villages in the Apennine Mountains, which run down the middle of the Italian peninsula. The districts of Novara and Pavia, for example, experienced an annual influx of 50,000 weeders. Coming from the highlands, these workers had little immunity to malaria. While working in the rice fields, they were exposed to conditions that fostered the transmission of the disease.[19] Although malaria in the North debilitated local populations, cooler temperatures largely prevented the transmission of *P. falciparum* malaria, and malaria mortality was low. Italians living in the South were not so fortunate, and there the nineteenth century witnessed extensive malaria mortality.

Malaria was probably endemic over much of southern Italy from Roman times. This may explain why the region remained sparsely settled following the great plagues of the fourteenth century, which depopulated much of Europe. Yet malaria was not the only reason for the region's sparse population. Land tenure rules from the fifteenth century were designed to encourage sheep raising and wool production to supply a growing textile industry. This led to low population densities. Feudal estate owners controlled large tracts of grazing land. In addition, the Bourbon regime, which gained control of the region at the beginning of the nineteenth century, discouraged the development of peasant agriculture in order to prevent the emergence of a large popular class of peasant farmers that might threaten the political stability of the region. Thus a feudal pastoral economy continued to dominate the region up through the middle of the nineteenth century.

Beginning in the 1840s, however, land in the South began to be bought up by an emerging class of capitalist farmers who sought to develop the region's agricultural potential. This process was accelerated following Italian unification in 1860. The new Liberal government sought to transform southern Italy by supporting the development of large agricultural estates owned by landlords who possessed the capital to increase the region's agricultural production. New land tenure rules encouraged the concentration of land in the hands of a small number of estate owners.[20] While small farms continued to exist in many parts of the South, the large estates dominated the region's economy, particularly in the rich low-lying agricultural lands.

Although the new landlords possessed the capital to acquire lands, they did not employ their resources to invest further in the modernization of agricultural production, as occurred in areas of northern Europe and the United States, as well as in the rice-growing areas of the American South. Instead the landlords sought to maximize profits by instituting an exploitive system of labor extraction. The landowners operated their estates through leaseholders, who worked on short-term contracts and had no vested interests in the long-term development of the land. They were concerned strictly with making a quick profit. The system of exploitation they promoted was the equivalent of "agricultural strip-mining" and created conditions that fostered the spread of malaria.[21]

To work the land, the leaseholders employed large numbers of peasant farmers and day laborers during the peak months of production in July and August. In Foggia, the great estates attracted 100,000 workers, many coming from the Apennine Mountains during the summer months. The population of the Roman Campagna increased by 30,000 each summer. In Apulia workers constituted a large army of landless workers who lived in what Frank Snowden described as "agro-towns," located above the agricultural plains.[22] In other areas, a variety of tenant relationships that provided peasant farmers with some hold on the land existed. Yet even in these areas peasant farmers required the seasonal employment of additional workers drawn from the hillside towns. Throughout southern Italy the population of potential farm workers greatly exceeded demand. Population growth during the second half of the nineteenth century, the concentration of land in the hands of a few landlords, and the absence of alternative forms of employment created a surplus supply of labor.[23] This imbalance gave leaseholders considerable leverage in defining terms of employment, often to the detriment of the health and well-being of the workers. Seasonal estate workers toiled long hours from dawn to dusk and were exposed to the bites of local malaria vectors during the early morning and evening hours. They were sheltered in stables and straw huts or camped out in open fields, further exposing them to malaria. Wages were low, and diets inadequate, reducing their ability to resist the effects of infections.[24]

The rise of the agricultural estates in the plains and hillsides of southern Italy during the second half of the nineteenth century also altered the region's physical landscape in ways that further fostered the spread of malaria. Large areas of the Apennine Mountains were deforested by the leaseholders trying to make a quick profit from planting on virgin soil and by displaced peasant farmers seeking a small piece of land to cultivate in the hills. The absence of deforestation laws facilitated these forest-clearing activities. Loss of forest cover exposed the land to heavy runoffs following annual spring rains. Torrents of water, no longer restrained by vegetation, swept away soil, set off landslides, and silted up river beds downstream. Fed in this manner, rivers and streams repeatedly overflowed, creating stagnant ponds in the valleys and along the coast, which, as so often happens, provided rich breeding grounds for local malaria-carrying mosquitoes. To this ecological disaster was added the failure of leaseholders to care for the lands under their charge, allowing these lands to remain waterlogged.[25]

The construction of new railway lines along both the Adriatic and Tyrrhenian coasts contributed further to the creation of breeding grounds for malaria mosquitoes. These railways were intended to promote the political unification of Italy and facilitate the movement of agricultural produce from the South. The embankments on which they were built, however, blocked drainage and created inland marshes similar to those resulting from efforts to hold back the sea in southeast England during the sixteenth century. Together these economic and ecological transformations led to the expansion of malaria in southern Italy at the same time that malaria was gradually disappearing from much of the rest of Europe.

Malaria did decline in southern Italy from the end of the nineteenth century. This decline has often been attributed to the development of a statewide system of quinine distribution that was initiated in 1900 (described in chapter 5). Malaria mortality, however, began declining as early as 1887 when statistics began to be kept. The decline in malaria mortality during the last two decades of the nineteenth century, like its rise from the 1860s, may be related to economic transformations in the region. Beginning in 1880 large estates began to be subdivided. Moreover, a depression in agricultural prices led some landlords to divest themselves of land. In addition, landownership became viewed by a growing middle class of lawyers, doctors, merchants, and public officials or *carabiniers* as important for their social advancement. Like earlier estate owners, this new class of entrepreneurial landowners chose to place the exploitation of their farmlands into the hands of leaseholders. Unlike the earlier estate owners, however, they engaged peasant farmers directly in various kinds of land tenure arrangements. As a result, day laborers increasingly became tenants, sharecroppers, or other kinds of *coloni*. This enabled them to begin accumulating small amounts of capital, which permitted some to buy a piece of land, build a house, and have some animals. But not all day laborers succeeded in improving their economic position in this way, and southern Italian peasants became increasingly stratified. Those who gained more direct access and control of land were able to improve their economic welfare, take better care of the land, and reduce their exposure to malaria. Those who could not afford to do so continued to work in poorer conditions and paid a price. Yet the acceleration of emigration after 1880 led to a decrease in the numbers of families in the poorest circumstances.[26]

Moreover, emigration reduced population pressures on local resources and revolutionized the labor market. Landowners, forced for the first time to compete for workers, chose to mechanize their farms, introducing steam-powered harvesters and threshers. As in the American South after 1935, these changes vastly reduced the numbers of seasonal workers employed out of doors at the height of the malarial season and shortened the workday of those required to carry out various agricultural duties. Shorter workdays allowed workers to avoid peak anopheline feeding times at dawn and dusk.[27] Malaria also declined in the North of Italy as a result of the mechanization of rice growing. In short, the capitalization of agriculture altered the human ecology of malaria, reducing exposure to the disease. Finally, migrants returning to Italy from the United States brought with them financial resources that allowed them to purchase land for farming and construct sturdy houses impervious to mosquitoes. The overall standard of living in many of the worst malaria regions improved. In effect, the economic transformations that had led to reductions in malaria in much of the rest of Europe began to occur in Italy. Despite these changes, however, malaria remained a problem in Italy until after World War II.

The persistence of malaria in the southern United States and Italy reinforces the importance that the conditions of agricultural production had for the epidemiology of malaria. Malaria remained a southern disease where forms of production placed human populations at risk of infection, and it began to disappear only when these conditions changed.

CHAPTER FOUR

Tropical Development and Malaria

Patterns of agricultural production in tropical and subtropical regions of Africa, Latin America, Asia, and the Pacific during the nineteenth and early twentieth centuries resembled those that occurred in Italy and the American South in a number of ways. They tended to be undercapitalized, used few technical inputs, relied on the extensive use of human labor, and were marked by significant inequalities in the ownership of land. Not surprisingly, they had similar consequences for the epidemiology of malaria. In many regions, patterns of agricultural development failed to provide producers and their families with the resources needed to improve their conditions of living or reduce their exposure to infection. It thus prevented them from "growing out of malaria." At the same time, in a number of places these patterns of production led to increases in the distribution and intensity of malaria transmission. While climate and the presence of highly efficient vectors contributed to the persistence of malaria, conditions of production played an equal if not greater role.

By the beginning of the nineteenth century, the industrialization of Europe and the United States was generating a growing demand for raw materials from around the globe. At the same time, European and American industries sought new markets for their industrial products. Over the course of the next 150 years, broad regions of the tropical and subtropical world were drawn into a rapidly expanding global economy, becoming increasingly dependent on the export of agricultural goods and the import of manufactured products. Reinforced by the parallel expansion of European and American imperial domination, and continued European hegemony in Latin America, the period witnessed a dramatic expansion of tropical and subtropical agricultural production.

Agricultural development in tropical and subtropical regions of the globe during the nineteenth and twentieth centuries followed two patterns, one based on smallholder production, the other on the creation of large plantations. Over much of Africa and South Asia, and some parts of Southeast Asia and Latin America, smallholder farming dominated agricultural production. While farmers in many of these regions had participated in market production before European colonial rule, colonial expansion drew these regions into a tighter relationship with global markets and increased demand for both existing and new agricultural commodities. Colonial governments employed a combination of incentives, including taxation, to encourage local farmers to grow cotton, rubber, cocoa, groundnuts, sugar, or other agricultural commodities for European and American markets.

Although the expansion of agricultural commodity markets created new opportunities for economic advancement in the tropics, the structures of colonial markets, as well as of those of newly independent states in Latin America, limited the ability of peasant farmers to profit from market participation. The existence of government marketing boards and other monopoly buying arrangements, designed to limit the cost of exported raw materials needed by European industries, and to increase the profits of commercial trading firms, kept the prices paid to local farmers low.

Agricultural policies that ensured redundancy in the distribution of sites of production had a similar impact. Thus the British, following the spike in cotton prices during the American Civil War, began encouraging cotton production wherever it could be grown within the empire. By doing so they successfully limited price increases that followed a decline in production in any one area due to bad weather, war, or disease. The policy was first applied in India during the 1870s and in wide areas of Africa during the first decades of the twentieth century. While this redundancy served the needs of British textile industries for cheap cotton, it kept the price paid to colonial cotton producers low. Portugal employed similar tactics, encouraging cotton production in its African colonies as well as in Brazil during the nineteenth century. The French also encouraged the widespread production of cotton, groundnuts, and rice. Low commodity prices prevented peasant farmers from accumulating capital that could have been reinvested into production. They thus restricted agricultural transformations that might have led to overall improvements in living standards.

Market production had additional drawbacks for peasant producers. Like their counterparts in the American South, tropical farmers who were drawn into the production of cash crops became vulnerable to fluctuations in international commodity prices. When economic recessions in the industrialized North led to a decline in demand for raw materials produced in the South, agricultural prices fell and farmers were often unable to make ends meet. Like cotton farmers in the American South, moreover, colonial farmers tended to respond to declining prices by expanding cash-crop production in order to maintain income levels. This often led to a reduction in the land devoted to growing subsistence crops and a growing dependence on cash-crop incomes for food. This dependence contributed to periodic famines, when declining cash-crop prices coincided with periods of extended drought and rising food prices.[1]

The consequences of expanding peasant participation in international commodity markets were not uniformly negative. A small proportion of farmers, with access to land and labor, and the ability to transport their crops to market, were in a position to take advantage of new opportunities. These farmers were able to expand their production and improve their standard of living. For most farmers, however, market participation provided few opportunities for advancement and often led to increased indebtedness and poverty. These conditions did not necessarily lead to an expansion of malaria in the tropics; however, they largely prevented colonial populations from "growing out of malaria." Poor housing and sanitation conditions continued to expose the vast majority of farmers living in malarious areas to infection.

The immiseration of poor farmers produced a continual flow of rural populations into emerging colonial cities in search of alternative sources of subsistence. Rural stratification had a similar impact in some parts of Europe and America, as we saw for example in southeastern England. Yet, unlike areas in the temperate North, expanding colonial participation in tropical and subtropical agricultural commodity markets was seldom paralleled by the growth of industry and industrial employment, because European powers viewed their tropical possessions as sources of raw materials and markets for industrial products manufactured in European cities. Where local industries emerged, such as the cotton and jute mills of Bombay and Calcutta, they employed only a small proportion of the landless or land-poor rural workers seeking employment.[2] Lines of unemployed gathered outside the cotton mills of Calcutta every morning in search of work during the early decades of the twentieth century.[3]

In the absence of industrial employment, the dispossessed in emerging colonial cities sought other forms of domestic or commercial employment or participated in the informal economy of piecemeal trading, prostitution, or crime. Low wages, poor housing, and the absence of sanitation measures marked conditions of living in the large slum areas, which grew up around colonial cities. In 1931 the *Report of the Royal Commission on Labour in India* noted that the housing of the industrial worker presented difficult problems, especially in cities like Bombay, where each small room of a tenement, three to four stories high, was occupied by at least one family and where insufficient space between the structures prevented the entry of light and air. Sanitary arrangements were said to be totally inadequate, and cleansing neglected. While new factories were constructing improved worker housing, many hundreds of workers remained homeless, living on the streets or verandas "under conditions of squalor."[4]

Urban locations, far from being a refuge, often became breeding grounds for malaria-carrying anopheline mosquitoes. The slum conditions of Bombay created multiple opportunities for *Anopheles stephensi,* which bred in any open collection of water, to reproduce and feed off human hosts.[5]

While small-scale peasant agricultural production dominated

colonial economies in many parts of Asia and Africa, large-scale plantations occurred with great frequency in the Caribbean and parts of Latin America, southern and eastern Africa, and parts of Southeast Asia. Tropical plantations produced a single crop, such as sugar, rubber, rice, tea, or cotton. Like the *latifundia* of southern Italy, production on colonial plantations relied on the use of large supplies of cheap seasonal labor, with only limited investments in mechanization or other capital inputs. While peasant-based production limited the ability of farmers to improve their resistance to disease, plantation economies often contributed to an expansion of malaria. They did this by creating breeding opportunities for malaria vectors, exposing susceptible populations to infection, and facilitating the movement of malaria parasites.

Plantation agriculture in the tropics introduced new patterns of land use as well as new forms of labor recruitment and control that transformed local ecologies and encouraged the transmission of malaria. In some regions, clearing of large tracts of land and eliminating tree cover changed the flow of water and created environments that were conducive to the breeding of new species of malaria-transmitting anopheline mosquitoes, at the same time that they eliminated nonmalarial species. In more arid regions of the tropics, plantation agriculture was made possible through the introduction of irrigation canals and impounded water. These innovations could facilitate the breeding of mosquitoes and the transmission of malaria. Moreover, in regions in which malaria had been seasonal, occurring in association with annual rains, irrigation extended the transmission season throughout the year. Irrigation, of course, did not necessarily increase conditions of exposure. Much depended on the habits of local anopheline mosquitoes and the susceptibility of the labor force.[6]

Tropical plantations frequently relied on seasonal migrant workers for their labor supply, and only a few workers were employed year round. The movement of workers back and forth between plantations and sourcing areas contributed to the expansion of malaria in several ways. Tropical plantations were often located in low-lying areas, which were susceptible to malaria. Where labor was available locally, the disease could produce some sickness but was seldom a serious threat to the work force, because local

populations were commonly resistant to the disease. Where labor
had to be recruited from some distance, however, malaria could
become a more serious threat to worker health. The use of labor
drawn from malarious areas could introduce new strains of the
disease into local populations, leading to upsurges in the disease.
Alternatively, many plantations found it necessary to import labor from regions in which malaria was either nonexistent or an
infrequent occurrence. Recruited workers from these areas had
little or no resistance to malaria and often experienced high levels
of morbidity and mortality, similar to those suffered by hill town
workers on the agricultural estates of southern Italy. In the eastern
Congo, for example, Belgian coffee estates located in low-lying
areas surrounding Lake Kivu recruited workers from neighboring highland areas of Rwanda during the 1930s. A survey of these
estates carried out in 1935 found extensive malaria among these
workers.[7] In the Zululand region of South Africa, the expansion
of sugar production from the beginning of the twentieth century
required the recruitment of labor from outside the region, because
local Zulu workers resisted working on the estates. The sugar industry attempted to recruit workers from regions that were similar
in climate and exposure to malaria; however, this proved difficult, and by the 1920s the industry was importing workers from
nonmalarious areas in Pondoland and Basutoland.[8] These workers suffered high rates of malaria morbidity during the late 1920s
and early 1930s. Similarly, during the nineteenth century, workers
migrating from the mountain highlands of Jamaica to work on
banana plantations along the coast were frequently exposed to malaria. Finally, large numbers of indentured servants were recruited
to work on the sugar estates in Malaya during the nineteenth and
early twentieth centuries. Most of these came from southern areas
of India where malaria occurred seasonally but not with sufficient
frequency to provide workers with protection from the disease
that they experienced in the highly endemic sugar growing areas
of Malaya.[9]

The role of migration in the spread of malaria among plantation workers was facilitated by the existence of poor working and
living conditions on many plantations. Tropical plantations were
notorious for their disregard for workers' health. Where hous-

ing was provided, it seldom included protection from invading insects. In some cases, housing lacked adequate ventilation, and workers preferred to sleep out of doors, where they were exposed to malarial vectors. In addition, little attention was paid to sanitation measures, and housing areas frequently provided breeding sites for local malarial vectors. Additional breeding locations were provided by poorly maintained irrigation systems. Plantations thus produced conditions that exposed populations to infection.

Not all employers of labor ignored conditions that contributed to malaria. Some large citrus estates in South Africa, for example, provided workers with housing and quinine to protect them from the disease. In the 1930s, the copper mining industry in what is now Zambia instituted major reforms to reduce malaria. Yet most tropical industries provided few protections.

Workers infected while working in malarious areas could infect those living in their home areas once they returned, contributing to the geographical expansion of malarial infections. In 1938 a highland epidemic of malaria, causing many deaths, occurred in one of the areas from which Belgians recruited Africans to work on coffee estates in the eastern Congo. An investigation found that the epidemic had been sparked by the return of workers who had been infected on the estates and in Uganda.[10] Malaria spread into other tropical highlands as a result of similar patterns of labor migration in New Guinea during the 1950s.[11] Finally, in Madagascar and the western highlands of Kenya, migrants were recruited from malarious lowlands to work on highland estates where malaria had previously not existed.[12] Predictably, malaria outbreaks occurred in the highlands when climatic conditions encouraged breeding of anopheline mosquitoes. In these ways, the labor practices associated with plantation agriculture contributed to the movement of malaria parasites into new areas.

An examination of three case studies can reveal how patterns of tropical development influenced the epidemiology of malaria during the first half of the twentieth century. In northeastern Brazil, the Punjab region of India, and South Africa, conditions of agricultural production increased the exposure of local populations to malaria infection, undermined their resistance to disease, and prevented them from accumulating the resources needed to grow out of malaria. These cases, while defined by local circumstances, are representative of conditions and historical processes that occurred with considerable frequency over wide areas of the tropical and subtropical South from the beginning of the nineteenth century.

PLANTATION AGRICULTURE IN NORTHEAST BRAZIL

In 1939 Rockefeller Foundation officer Frederick Soper submitted a report of his effort to eradicate the *Anopheles gambiae* mosquito from Brazil. The highly efficient malaria vector had accidentally been transported to the New World from West Africa by ship in 1931. It had subsequently gained a foothold in the region, contributing to a series of malaria epidemics. The most serious of these began in 1938 in northeastern Brazil. The epidemic started on the coast but quickly spread up river valleys into the interior, where it took the lives of thousands of people.[13]

Buried in Soper's report was a photograph labeled, "In time of drought thousands of families migrate to the coast or the Crato Valley." There is no other mention of the photograph or its significance to Soper's gambiae eradication campaign. Yet the photograph provides a window into the world of poverty within which a large portion of the population of northeast Brazil lived. This world played an important role in the human ecology of malaria and in driving the epidemic that Soper confronted. Examining who these people were and why they were heading to the coast reveals the ways in which patterns of tropical agricultural development contributed to the persistence of malaria in the tropical South.

To understand who these migrants were we need to look briefly at the ecology and economy of northeast Brazil in the 1930s. Like much of Latin America, landownership in northeast Brazil was concentrated in the hands of a relatively small number of estate owners. Many of these estates dated from the early Portuguese conquest of northeast Brazil in the seventeenth century. The estates were distributed over three ecological zones. The first, the Zona da Mata, consisted of a thin coastal plain that ran along the region's southeast coast. Here rainfall levels and soil characteristics supported large sugar plantations. Other crops, including cotton, were grown in areas unsuitable for sugarcane. Labor for

Refugees Fleeing Drought, 1937–1938
Source: Frederick L. Soper, *Anopheles Gambiae Eradication Campaign, Annual Report, 1939*, Rockefeller Archives Center, Sleepy Hollow, New York, RG5, series 3, box 113, photo 2138 rrr.

the sugar estates came from several sources: squatters who lived on the lands owned by the sugar estate owners, providing labor in return for land to grow subsistence crops; outside laborers from towns and villages of the zone; and, most importantly, seasonal migrant laborers from the interior. Up to 80 percent of the labor force at harvesttime from December through February was made up of these migrants. When the first rains appeared these migrants would return to their homes in the interior.[14]

To the west of the Zona da Mata was the Agreste, a transitional zone between the humid coastal plain and the dry interior regions that make up the vast majority of northeast Brazil. Here, as in the Zona da Mata, landholding during the first half of the twentieth century was concentrated in relatively few hands, with the large proportion of people living as labor tenants. Many Agreste squatters routinely augmented their income through migrant labor to the coast.

To the northeast of the Zona da Mata and the Agreste was a

1938–40 Gambiae Epidemic
Malaria High Risk Area 2000

Malaria in Brazil

vast arid region known as the Sertão. The Sertão encompassed nearly all of the states of Ceara and Rio Grande do Norte where the 1938–39 malaria epidemic occurred. The primary economic activity in much of the Sertão was cattle raising. However, agriculture played some role, particularly in the river valleys that transected the Sertão running from west to east. During the rainy season, these valleys flooded. As the rains receded during the dry season, the rivers dried up leaving silt-laden riverbeds that provided land for growing manioc, maize, and beans along with some cotton and sugar production. They also left behind pools of water that would provide an attractive breeding ground for the invading *A. gambiae* mosquito.

As in the Agreste and Zona da Mata, a few landowners, whose

families had received land grants in the eighteenth century, controlled the land in the Sertão. Each grant consisted of one or more long narrow estates (*fazenda*) stretching away from the region's major river valleys. The majority of families living on these estates were landless farmers working as sharecroppers, labor tenants, or renters. Sharecropping was the most popular form of labor relationship on the *fazenda*. As was the case in the American South, the owners furnished the land and seed, financing the peasant during the growing season. After the harvest, the sharecropper paid back his debt by providing one-half the cotton crop to the landowner. If the season was a poor one, however, the proportion a sharecropper had to pay might increase in order to cover his debts. The sharecropper retained the rest of the cotton crop and all of the maize and other food crops. In addition to sharing the crops, tenants were required to provide one day a week in labor to the landowner. Away from the river valleys, squatters raised cattle for the landowners.[15]

The system of sharecropping provided Sertão peasants with a fragile economic base, which left little room for accumulating even small amounts of capital. To make matters worse, conditions of production were changing on the estates of the Sertão during the 1930s. In order to compensate for the fall in agricultural prices brought on by the worldwide depression of the early 1930s, landowners sought to reduce their production costs by requiring their tenant farmers to provide more direct labor to the landlord. This increase in labor demands reduced the amount of labor that the tenant could devote to his own crops and thus his income.

Sharecropper and herder families supplemented their meager incomes by sending members of their families to work as seasonal laborers in the agricultural areas of the Agreste and Zona da Mata. Wages paid to seasonal workers were low and, in the late thirties, were undermined by rising inflation rates.[16] Yet migrants had little choice. Even in the best of times, working on the coast could mean the difference between a Sertão family surviving or not.

The lives of peasants and herders, difficult in normal years, were severely threatened by years of drought. Drought was a regular event within the region, occurring with amazing frequency throughout the nineteenth and twentieth centuries. Rainfall records for Ceara from 1913 through 1941 indicate the occurrence of below normal rainfalls every 5 or 6 years. Soper indicated that severe and prolonged droughts occurred approximately every 11 years. Drought undermined the ability of sharecroppers to meet the debts incurred from the landowners before harvest and, if sustained over two or more seasons, could lead to eviction. It also resulted in a costly loss of cattle.[17]

For the peasants and herders of the Sertão, who were living on a thin margin of subsistence, with little in the way of a safety net, the primary response to drought was increased migration. During extended droughts, Sertão peasants migrated by the thousands to the coast or more well-watered river valleys in search of food and employment.[18] Sertão men also traveled west to the Amazon region to earn income collecting rubber. These responses to drought played an important role in the epidemiology of the 1938–39 epidemic.

Records collected by Brazilian health officials indicate that malaria occurred seasonally over large areas of northeast Brazil, though transmission was more intense along the wetter coast than in the drier interior. Malaria was also intense in the Amazon region. These differences in transmission levels produced moderate levels of acquired resistance in coastal and Amazon residents but little or none among the residents of the Sertão.

Sertão migrants who traveled to the Amazon or coast during periods of drought were particularly susceptible to malaria. Frequently, they returned home infected with malaria parasites and set off local outbreaks. One occurred in 1934–37 following the return of migrants who had traveled to the Amazon in response to a three-year drought that began in 1932. When drought commenced again in 1936, men once more sought income in the Amazon. Their return in 1938 contributed to the 1938–39 epidemic in the Sertão. However, refugees returning from the coast were more important contributors to this epidemic.

The extended drought that began in 1936 drove large numbers of Sertão families to flee to the coast in 1937 and 1938. These migrant-refugees, with little or no resistance to malaria and suffering from hunger, malnutrition, and perhaps other infections, were hard hit by the epidemic, which, aided by the presence of

A. gambiae mosquitoes, had spread from the coast into the lower Jaguaribe River Valley. These refugees probably made up a sizable proportion of those who died during the 1938–39 epidemic. Those who survived fled the epidemic, returning to the Sertão, where they added to the reservoir of infection begun by returning Amazon migrants. As Soper noted in his annual report, "The emigration from the malaria ridden lower valley brought untold numbers of gametocyte carriers to the newly infected areas."[19] The convergence of these two streams of infection set the stage for a major epidemic across the Sertão. The arrival of the *A. gambiae* mosquito had the effect of gasoline thrown on a fire. Officially some 5,000 people died from malaria during 1938 and 1939. Unofficial counts were much higher. According to one local paper, "The human language is far from adequate to describe the desolation which existed in the region, in which suffering, tears and mourning spread their lugubrious mantle over thousands of graves. The general belief was that the Northeast would be depopulated because those who did not die at once would abandon it."[20]

Although the *Anopheles gambiae* played an important role in determining the severity of this epidemic, the human ecology of malaria in the Sertão was shaped by patterns of economic stratification, poverty, and migration that were part of the day-to-day life in northeast Brazil during the 1930s. These conditions exposed populations to infection and contributed to the movement of malaria parasites.

PEASANT FARMERS AND IRRIGATION IN THE PUNJAB

In the autumn of 1908, a catastrophic epidemic of malaria tore through the Punjab region of India causing an appalling loss of life. In October and November alone, the disease took 300,000 lives. For the year as a whole, more than 1 million people perished from malaria. The 1908 epidemic was the crest of a wave of malaria epidemics that had caused hundreds of thousands of deaths each year for the previous 30 years. These epidemics occurred in large measure because the region's normally dry conditions prevented malaria transmission for most of the year, and the population developed little immunity to the disease. When the annual monsoon flooded the region, creating conditions that for a short period of

South Asia

time enabled the widespread breeding of anopheline mosquitoes and the subsequent transmission of malaria, the population had little defense against the ensuing onslaught of disease.

In 1911 the government of British India published its official report on the causes of the 1908 epidemic.[21] Major S. R. Christophers, one of the leading malariologists in British India, wrote the report. As part of his inquiry, Christophers examined the history of malaria epidemics in the Punjab over the previous 40 years. To no one's surprise, his detailed report pointed to the important role of rainfall in determining the severity of the annual epidemics in the Punjab. Yet Christophers also noted that the observed variability in mortality over the period examined could not be explained by rainfall alone. Economic stress and hunger, reflected in rising grain prices in the years preceding epidemics, were important "human factors" in determining the severity of Punjab epidemics, and starvation contributed to the mortality caused by malaria.

Public health officials in India accepted Christophers's conclusions regarding the important role played by hunger in the epidemiology of malaria, and grain prices were used to predict epidemics over the next half century. His work has fared less well among more contemporary observers, who have questioned his underlying premise that hunger made individuals susceptible to malaria. To the contrary, some studies have shown a negative correlation between hunger and malaria. Historian Sheila Zurbrigg revisited this debate, reexamining Christophers's Punjab data using current statistical methods. Zurbrigg's regression analysis confirmed his finding regarding the role of both rainfall and grain prices, though the importance of grain-price variations was somewhat less than that proposed by Christophers. Regarding the role of severe hunger in malaria mortality, Zurbrigg rightly noted that critics of the role of hunger in malaria mortality had confused disease severity with lethality. While hunger may have a protective effect in limiting the ability of the malaria parasite to infect red blood cells, "acute starvation is associated with severe metabolic imbalance, making a person exquisitely vulnerable to additional physical stress." At the same time, immunocompetence was likely to be compromised, affecting a person's recovery rate. In short, the imposition of malaria on top of acute starvation increased the likelihood that an individual infected with malaria would die from the disease.[22]

But neither Christophers nor Zurbrigg addressed the reason why food shortages and famine stalked the lives of the Punjab's poor. While it is certainly true that periods of drought reduced food supplies and led to hunger, it is also true, as Amartya Sen and others have shown, that weather conditions were but one of the factors that determined the availability of grain—and thus the extent of hunger in India and elsewhere. The structure of agricultural markets, the existence or absence of social safety nets such as famine stores, and the availability of cash to purchase food all played a role in determining whether an individual or family could acquire food during times of scarcity and thus if drought turned into famine.[23] To understand why food shortages plagued the people of the Punjab and contributed to annual malaria epidemics, we need to look at the history of agricultural development in the southeastern districts where the correlation between grain prices and malaria was the strongest.

At the beginning of the nineteenth century the peoples inhabiting the districts of southeast Punjab lived in the shadow of drought and under the threat of famine. Agriculture depended on the existence of adequate rainfall, and dry years brought hunger. Local reliance on uncertain rainfall had led former rulers of the region to construct irrigation canals, the oldest being the Western Jumma Canal, which extended from the western side of the Jumma River. The canal dated from before the sixteenth century but had fallen into disrepair. When the region came under British control in the nineteenth century, efforts were made to remodel the canal in order to bring irrigation to the region.

The British oversaw the development of a wide network of irrigation canals over formerly arid regions of the Punjab during the second half of the nineteenth century. In the northern and central areas of Punjab, these canals were intended to provide the basis for agricultural colonies that would adopt modern cultivation methods. These communities would revitalize Punjab's agricultural economy and provide an outlet for surplus populations living in surrounding agricultural communities that were viewed as overcrowded and land starved. To ensure that these colonies would become model agricultural communities, the British retained ownership of the land and tried to enforce agricultural regulations.[24]

In southeastern Punjab, by contrast, the British canal rehabilitation efforts had less to do with generating development than with generating revenue. They simply reconstructed the canals and then charged water fees to those who drew water from them for cultivation. No controls were exercised over ownership of the land, and no efforts were made to shape the patterns of agricultural development that followed on the opening up of the canals. Instead, the British established a system of land rents in the Punjab that encouraged wealthy capitalist farmers and moneylenders

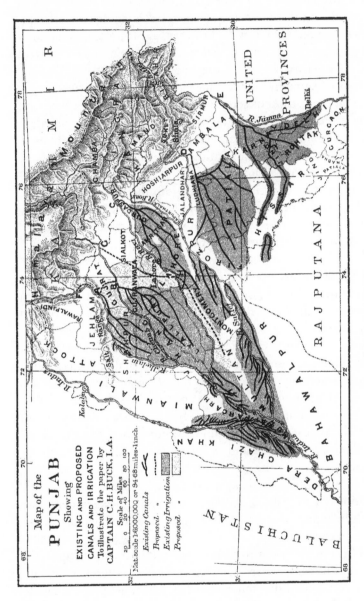

Punjab Irrigation Canals, 1901

Source: C. H. Buck, "Canal Irrigation in Punjab," *Geographical Journal* 27, 1 (January 1906): 61.

to buy up land and rent it out to sharecroppers or to employ agricultural laborers. Despite this lack of concern for generating development, the canals attracted large numbers of cultivators from neighboring areas and led to a major extension of cultivation in the region.

Yet the canal system and participation in irrigation farming came at a price. Most directly, both landowners and the cultivators who worked the land were required to pay water fees. The fees were based on the value of the crops planted and were fixed. Thus fluctuations in the value of the crops actually grown had no effect on the fees charged. When agricultural prices were low, the rates became a severe financial burden. Many cultivators found themselves seeking credit from landlords and other moneylenders in order to pay their water fees. Over time, poorer farmers experienced increasing debt burdens. Some of those who owned land were forced to sell it. In addition, many farmers adopted agricultural practices aimed at reaping short-term profits in order to reduce their debt and increase their income. These practices included shifting from subsistence cultivation to the growing of higher-value export crops and especially cotton, sugarcane, and wheat; overcropping, or the planting of crops year after year without allowing fields to go fallow; and extending irrigation to cover more acreage. All three practices led to short-term increases in income, but they all had long-term negative consequences for the farmers' ability to survive.[25]

The shift to cash crops increased the farmers' dependence on agricultural markets for their food. This made them more vulnerable to changes in food prices. Overcropping quickly led to the exhaustion of farmland, greatly reducing its productive capacity and the farmers' ability to earn the income needed to purchase food. Finally, the expansion of irrigation led in many cases to the oversaturation of the soil. As a result, many areas became waterlogged. In addition, evaporation of overwatered lands drew salts from the soil to the surface. This saline effervescence impeded the cultivation of crops.

Heavy dependence on purchased food combined with declining farm income, which resulted from overcropping and the heavy saturation and salination of the soil, explains why rising grain prices frequently led to starvation in southeastern Punjab. It also explains why there was a higher correlation between rising food prices and malaria mortality in this region of the Punjab than elsewhere. Increasingly destitute farmers faced starvation when drought led to sharply rising grain prices. Starving farmers and their families in turn suffered increased mortality from malaria when drought was followed by the autumnal epidemic of malaria. Thus, the particular form of agricultural expansion initiated under British rule laid the groundwork for increasingly severe epidemics of malaria in southeast Punjab.

LAND ALIENATION IN SOUTHEASTERN AFRICA

By the nineteenth century malaria was endemic over most of tropical Africa south of the Sahara. Only in the highland areas in eastern and southern Africa, dry areas of the Sahel and Kalahari Desert, and the temperate regions of South Africa was malaria not a part of daily life. Yet, even in some of these regions, epidemic malaria probably occurred occasionally where climatic conditions permitted an expansion of the range of malaria vectors.[26]

It is difficult to know to what extent colonial development policies altered transmission patterns in areas in which malaria was already holoendemic. The expansion of irrigation farming may well have contributed to increases in malaria in some areas of Africa as it did in India. Critics of the large-scale irrigation projects launched by the French in West Africa and by the British in Sudan argued that both contributed to increases in malaria. But the record is by no means clear.[27] What is evident is that colonial efforts to expand agricultural production did little to reduce the transmission of malaria in these regions. It is also evident that colonial policies contributed to the expansion of malaria into areas in which it had not previously existed or had occurred infrequently. Highland areas of the eastern Congo, Rwanda, southern Africa, the western highlands of Kenya, and the central highlands of Madagascar, all experienced increases in malaria transmission related to changing patterns of agricultural production during the colonial period. The processes leading to malaria expansion in Africa can be seen more clearly by looking at the history of malaria in the eastern and northern lowveld regions of present-day South Africa.

The Drakenberg and Wittenberg Mountains, which run parallel to the Indian Ocean in a great curve from Cape Town in the south to the northeast corner of the country, dominate the geography of South Africa. Between the mountains and the ocean is a lowland coastal plain, known locally as the lowveld. To the north and west of the mountains lies a vast highland plateau, or highveld, that covers nearly two-thirds of the country.

Historically, malaria has been confined to the lowveld north of the 30th parallel. South of this line, a combination of cooler temperatures and lack of rainfall discouraged the reproduction of malarial parasites. Cooler temperatures also inhibited the expansion of malaria onto the highveld plateau. Beginning in the 1920s, however, a series of economic policies instituted by the white-settler-dominated government contributed to both an expansion of malaria in the eastern and northeastern lowveld and its extension into the previously malaria-free highveld.

In the eastern and northeastern lowveld, malaria may have been present for centuries. However, it appears to have become a greater problem during the nineteenth century. While archaeological data support the existence of substantial agricultural communities in the lowveld at the beginning of the nineteenth century, Africans had abandoned much of the eastern lowveld by the end of the century, choosing to locate their homesteads on higher ground above the valleys of the De Kaap, Queens, Crocodile, Komati, and Lomati Rivers. These farmers continued to plant their fields in river valleys in order to acquire water for irrigation. However, they preferred to establish their kraals on higher ground above their fields. African testimony indicated that their preference for higher ground was dictated largely by the presence of disease in the valley.[28] The higher regions were also easier to defend.

The experience of early European settlers in the eastern lowveld in the 1840s attested to the severity of lowveld malaria. The town of Ohrigstad was founded in 1845 under the leadership of A. H. Potgeiter. During the first summer after their arrival, a few settlers fell ill and died of fever. A year or two later the losses were much greater. An old walled cemetery with some 25 recognizable graves, testified to the four tragic years that the settlement survived.[29] In 1848 the settlement was abandoned, and Lydenburg was es-

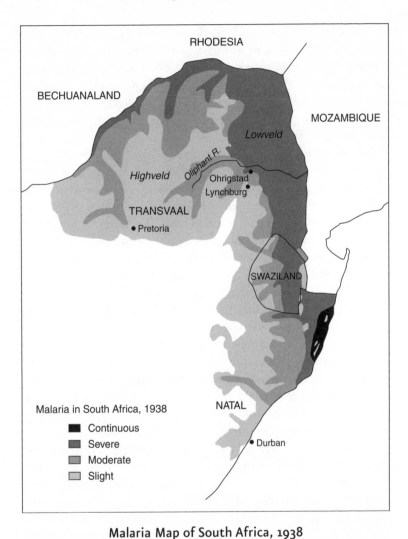

Malaria Map of South Africa, 1938

Source: David le Sueur, Brian L. Sharp, and Chris Appelton, "Historical Perspective of the Malaria Problem in Natal with Emphasis on the Period 1928–1932," *South African Science Journal* 89 (May 1993): 233.

tablished further south and at a higher altitude.[30] While malaria continued to take its toll in Lydenburg, which means "town of suffering," the effects were less severe than in the lower reaches of the district. Europeans, like Africans, largely avoided the lowveld until the 1920s.

The risks of exposure to lowveld malaria for Africans was dramatically demonstrated in 1896, when people from upland areas were recruited to help in the construction of a railway line connecting the Transvaal with the Indian Ocean port of Laurenco Marques (Moputo) in Portuguese East Africa (Mozambique). According to the railway medical officer, Dr. H. A. Spencer, malaria tore through the construction camps taking many lives.[31] From the information provided by Spencer, it would appear that the epidemic was sparked by the presence of the nonimmune Africans from the highlands working and living in crowded lowland camps, which produced ample breeding sites for *A. gambiae* mosquitoes. These workers experienced an acute infection, probably caused by *P. falciparum* malaria, which produced many deaths. Some of those who survived and returned home evidently carried parasites with them, providing a reservoir of infection from which the disease spread into the highveld.[32]

The experience of these railway workers was repeated on a much broader scale in the 1920s and 1930s when white settlers evicted large numbers of Africans from their hillside homes. The higher regions in which lowveld Africans had settled at the end of the nineteenth century had been officially designated for white settlement during the initial period of white occupation and conquest of the Transvaal, beginning in the 1840s. However, few whites had actually settled there because these areas offered limited economic opportunities and were regarded as malarious. This situation changed in the 1920s, when an expanding market for citrus crops led whites to plant orchards in the lowveld. These white farmers came to share African preferences for living on the higher areas adjacent to river valleys in order to avoid the worst effects of malaria.

In the 1930s a second wave of whites moved into the region. These were largely poor whites fleeing the Depression. Many were retrenched workers or farmers who had lost their land in higher districts. As a group, they sought new land at cheap prices or employment with more well-to-do white farmers.

The movement of white settlers into the region forced African farmers to choose between becoming labor tenants for a white landowner and moving to areas that had been designated for African settlement. These African reserves were located primarily in areas in the lowveld that were known to be malarious.[33] Despite the threat of malaria, many chose to move down into the lowveld rather than become subservient to a white farmer. The African populations of the lowveld districts expanded during the 1920s and 1930s by roughly 50 percent. Most of this expansion occurred in the lower, more malarious parts of the district and resulted from the movement of Africans out of higher areas of the country.[34]

Those African farmers who chose to move into the lowveld with their families faced an increased threat from malaria. It is difficult to measure the impact that this movement had on the health of African farmers. Few statistics were kept on African sickness or deaths. However, something of this experience can be gleaned from the testimonies of local white residents, missionaries, and magistrates. These records include a number of cases in which Africans who moved to lower areas suffered severe loss of life from malaria. In 1930, for example, roughly 1,000 Africans were forcefully removed from their highland farms by the forest department. They moved to an adjacent farm located 1,000 feet below. More than 70 of these displaced Africans died from malaria during the following six months.[35]

Although Africans moved to the reserve areas of the lowveld to acquire land, the quality of reserve land was low, and many lowveld Africans were forced to subsist by working on the white-owned citrus farms that were established in the area during the 1920s. Working and housing conditions on most white farms were poor. Smaller farms, often operating on a thin margin of profit, provided no screening for their workers' quarters. Nor did they invest in actions that would eliminate breeding sites near the worker quarters. Some farmers even refused to provide their workers with quinine tablets during malaria outbreaks or, if they did, required the workers to supply additional labor in exchange for quinine tablets.[36]

The neglect of worker health was fairly pervasive in southern Africa during this period as long as labor supplies were sufficient. Yet the specific neglect of malaria protection was based on the mistaken belief, promoted by public health authorities, that Africans living in the lowveld possessed immunity to the disease. The real-

ity was that most of the lowveld African population had recently
settled in the region and had not experienced long-term exposure
to the disease. As a result , they had little immunity to it, and annual outbreaks of malaria occurred between February and June.
Even worse, African populations in the lowveld were hit by a series
of particularly severe malaria epidemics in 1923, 1926, 1928, 1936,
and 1942.

Whites also suffered from malaria in the lowveld, but there was
considerable variation in their susceptibility. Well-to-do whites
were able to invest in screening and the regular use of quinine
during the malaria season. As a result, they experienced little malaria, even in the most highly endemic areas. By contrast, poor
whites, of which there were an increasing number from the 1930s,
were unable to afford these precautions and suffered considerable
mortality from malaria before World War II. Thus infection was
by no means inevitable for whites living in malarious areas but was
instead largely determined by class.[37]

Although malaria was generally limited to the lowveld during
the first three decades of the twentieth century, it began making regular incursions into the neighboring highlands during the
1930s. This occurred in Swaziland in 1932 and in the eastern Transvaal in 1937 and again in 1939.[38] In each case, malaria expanded
into areas where it had seldom, if ever, occurred. In each case as
well, its spread was facilitated by colonial development policies.

The history of the 1939 Transvaal epidemic reveals a great deal
about the ways in which patterns of social and economic development interacted with climatic factors to produce the periodic
expansion of malaria into highland areas of southern Africa. To
understand the origins of the 1939 epidemic, it is important to
look at the terrain in which it occurred and the populations it affected.

The geography of the eastern Transvaal was and is marked by
a number of river valleys that descend from the highveld to the
lowveld. The largest of these valleys is created by the Oliphants
River, the headwaters of which are located near the capitol city of
Pretoria, 4,364 feet above sea level. From there the river descends
gradually to the northeast, carving out an increasingly wider and
deeper valley as it makes its way toward the lowveld. The Croc-

odile, Sabie, Steelpoort, and Komati Rivers created somewhat smaller valleys, each of which provided an avenue for the periodic expansion of malaria into the highlands. But it was the Oliphants River Valley that seems to have provided the main conduit for the spread of the 1939 epidemic.

A. funestus mosquitoes bred annually along the banks of the lower reaches of the Oliphants River, contributing to moderate amounts of malaria each year. The presence of malaria made much of the lower valley unattractive for white settlement, and most of the area was included in the Sekhukhuneland Native Reserve. During years marked by excessive rains, flooding occurred, followed by receding waters, which left behind small pools. These pools were attractive breeding sites for a more efficient malaria transmitter, *A. gambiae.* As a result, malaria transmission intensified. In 1939, in the face of a severe outbreak, many African families living near the river abandoned their homes and sought shelter in the higher areas of the reserves. In doing so, they introduced malarial parasites into the upland areas.

The 1939 epidemic did not contain itself to the Sekhukhuneland Reserve, though it took a high toll there. Instead it spread out over wide areas of the eastern Transvaal highlands, affecting a large number of white residents as well as Africans. A series of additional factors fed this wider epidemic. To begin with, many Sekhukhuneland residents regularly sought employment on the white farms that were scattered over the middleveld and highveld areas of the Transvaal. Those infected with malaria may have contributed to the outward spread of the disease. White residents of the highlands certainly pointed to the reserves as the primary source of infection in the wake of both the 1937 and 1939 epidemics. Yet other factors likely played more important roles in these epidemics.

Many of the whites living in the so-called bushveld region bordering the Sekhukhuneland and the Oliphants Valley were tenant farmers or employees of the Forest or Irrigation Departments. Many had lost their farms during the Depression. They were often poor and, like their counterparts in the southern United States, lived in poorly built houses, without sanitation and with little protection from malaria mosquitoes. A health inspector in the

Waterberg area located to the west of the Oliphants River Valley provided a description of the housing conditions on white farms: "Very few farmers in this district thoroughly mosquito proof their houses. . . . In most cases the farmer is not the registered owner of the farm he lives on, and is therefore not willing to build on the property of someone else. He therefore erects a hovel of unburnt bricks or wattle and daub, thatches the roof and this shack serves as a temporary home sometimes for years on and costs: Nothing!"[39] Concerns about the declining health of poor whites following the Depression led the South African government to create a Rural Rehabilitation Scheme. The scheme provided poor whites with funds to improve their houses. The scheme was withdrawn in 1937 on the assumption that it was no longer needed, despite the fact that many poor whites continued to live in substandard housing with little or no protection from invading mosquitoes, and it was reinstated in the wake of the 1939 epidemic.

The Rehabilitation Scheme had another benefit: it provided funds for building dams to provide water for irrigating white farms. During the 1930s, streams and small rivers throughout the bushveld were dammed, creating small ponds. Unfortunately, while these ponds supplied local residents with water, they also provided a chain of breeding sites that facilitated the spread of *A. gambiae* mosquitoes from the lowland breeding areas up onto the higher areas of the country.[40]

It is highly likely that the further spread of malaria into highland areas as far away as Pretoria resulted from the movement of infected poor whites either fleeing the epidemic or seeking care from relatives. In the end, health authorities reported 9,300 deaths due to malaria during the epidemic—unofficial counts ran as high as 20,000.

❖ ❖ ❖

The history of agricultural development in the American South, Italy, and tropical and subtropical areas of Asia, Africa, and Latin America during the nineteenth and early twentieth centuries helps to explain why malaria, by the first half of the twentieth century, had become a tropical disease. Patterns of economic development linked to agriculture expansion prevented farmers in these regions from growing out of malaria at the same time that they contributed to an extension of malaria, by creating conditions that fostered the breeding of anopheline mosquitoes, exposed populations to malaria infection, and contributed to the movement of malaria parasites. As we will see in chapter 7, similar patterns of agricultural development continued to shape the epidemiology of malaria in the world's South during the second half of the twentieth century. Yet, from the end of the nineteenth century, it was precisely these societal forces that were systematically eliminated from discussions of malaria, and malaria became a disease of parasites and mosquitoes. In the next chapter we explore why this happened.

CHAPTER FIVE

The Making of a
Vector-Borne Disease

In 1900 Italian malariologist Angelo Celli published *Malaria: According to the New Researches,* in which he described the epidemiology of malaria and its prevention based on the recent discovery of the malaria parasite and the role of anopheline mosquitoes in malaria transmission. Celli's views of both the cause and prevention of malaria were not limited to the biological relationship among malaria parasites, anopheline mosquitoes, and human populations. For Celli, malaria was "connected with the economic and political life of the people who inhabit the regions where it dominates." Thus, looking back at the history of malaria in Europe, he observed how the disease had largely disappeared from northern Europe but persisted in the South and concluded that this disparity was related to different patterns of economic development: "In Northern Europe, in England, in Germany and in France, as in Northern and Central Italy, examples of vast and successful schemes of agrarian sanitation are numerous, while unfortunately not one can be cited from Rome to the most southern parts of Italy, where for centuries no marked permanent amelioration in the prevalence of this epidemy has taken place."[1]

Celli's book contained extensive discussions of the social conditions that contributed to the intensity of malaria in southern

Italy, including the exploitation of hill town workers forced to toil under unhealthy conditions on lowland estates where they were exposed to malaria infections. As a public health official, Celli supported the widespread distribution of quinine as a means of reducing malaria mortality. However, he did not believe that quinine alone would solve Italy's malaria problem. Instead he called for the abolition of the system of agrarian feudalism by which "the land remained in the hands of a few, who had not, owing to self-interest or to the dislike of innovation, any desire to change the ancient system of cultivation."[2] The elimination of malaria, in short, required social reform and effective agrarian development.

Although Celli was a radical with sympathies for the political left and the trade-union movement, he was not alone in linking malaria control to economic development. One of Britain's leading malariologists, Colonel S. P. James, repeatedly argued that malaria control would follow agricultural development. Professor M. A. Barber, a noted bacteriologist who worked for the U.S. Public Health Service on the control of malaria in the American South during the 1920s, also believed that the elimination of malaria would follow general social and economic advance. Barber, who was elected president of the American Society of Tropical Medicine in 1940, developed the use of Paris Green, an extremely poisonous bright green powder prepared from arsenic trioxide and copper acetate, which proved effective in attacking the larval stage of anopheline mosquitoes. Yet, like Celli, he held a broader view of malaria and its control. Writing about malaria in the rice-growing areas of Louisiana and Arkansas, he observed:

> Apparently the amelioration in malaria prevalence in *Anopheles* infested regions occurs only where improved agriculture or other social factor has brought about a general economic improvement in the condition of people. . . . It would seem that with even a moderate betterment of social conditions, malaria in the United States tends to disappear or become relatively inconsiderable provided such improvement is general. Or, to state the proposition in another way, the maintenance of high endemic malaria requires a permanent reservoir of infection such as is furnished

by a considerable body of people lacking proper housing, proper food, and adequate medical treatment.[3]

In fact, many of the most prominent malariologists in the twentieth century acknowledged that broader social and economic forces shaped the epidemiology of malaria and that the effective control of malaria required simultaneous advances in social and economic development. Yet efforts to control malaria from the end of the nineteenth century seldom mirrored these broader views. Instead, they reflected a fascination with the biometrics of malaria, as it became understood during the second half of the nineteenth century and the first decades of the twentieth century. Efforts to control malaria relied increasingly on narrow biomedical solutions, culminating in the 1950s in the adoption of long-acting pesticides and antimalarial drugs, and more recently in the widespread use of insecticide-treated bed nets. Each of these technologies has been effective in limiting or preventing malaria transmission and in saving lives. Yet neither individually nor collectively have they succeeded in eliminating malaria as a global health problem. As we will see in chapter 7, the problem has become worse in many parts of the globe over the past 30 years. The limited effectiveness of recent efforts to eliminate malaria as a public health problem stems in large measure from a failure to appreciate the importance of social and economic forces in driving the epidemiology of the disease. Or, put another way, it flows from a failure to appreciate the lessons of history. The present chapter examines how malaria control became a problem of drugs, nets, and insecticides during the twentieth century.

Human populations have often found ways to cope with the threat of malaria. As we saw in the preceding chapters, the primary response was often avoidance. Communities came to recognize and evade places where malarial fevers occurred, even if they did not understand why those places were malarious. They would locate their settlements away from swamps and marshy areas or avoid them during those periods when malaria was at its height. While it was not until the end of the nineteenth century that European scientists began to unravel the complex life cycle of malaria parasites and the role of anopheline mosquitoes in malaria transmission, there are numerous references to the relationship of mosquito bites to fever in medical writings from ancient Rome, India, and China. Avoidance, of course, was not always an option. Sharecroppers in the American South, slaves working on agricultural estates in Sardinia and in the Roman Campagna, and hill town residents farming the estates of southern Italy had little choice about when and where they worked, and they were repeatedly exposed to malaria.

Populations exposed to malaria also developed knowledge of medications that could reduce its impact. Recent *in vitro* studies of plants used by traditional healers in various parts of the world to treat fevers have revealed that many exhibit strong antimalarial properties.[4] Hundreds of such drugs have been identified in various parts of Africa, Asia, and Latin America where malaria is prevalent. Most recently, Western biomedical researchers have recognized the antimalarial properties of the Chinese herb *Artemisia annua*. First described in Chinese medical texts dating from 2,000 years ago, it is the source of artemisinin, which is highly effective in treating falciparum malaria. Certainly the best-known plant to have antimalarial properties was the bark of the cinchona tree, first observed by Jesuit priests in Peru in the early seventeenth century. Cinchona, however, was not used by local populations to treat fevers. Its muscle-relaxing properties reduced the shaking effects caused by severe chills in the Andes. The Jesuits applied a drug made from cinchona bark to patients suffering from "fevers and ague" in hopes of reducing the severity of the chill phase of the disease. It proved, however, to have additional curative properties and became highly prized in Europe during the seventeenth century. In the early nineteenth century scientists were able to extract its active ingredient, quinine, which became widely used throughout regions of European settlement to combat malaria. Later research showed that quinine attacks the merozoite stage of the malaria parasite circulating in the human bloodstream. Like avoidance, the use of various antimalarial drugs occurred without direct knowledge of the malaria parasite on which they acted.

Recognition of the relationship between marshy regions and fever also led to efforts to drain swamps and reclaim land for cultivation. These projects, in the absence of knowledge of the role of anopheline mosquitoes in the transmission of malaria, had mixed results. A great deal depended on the nature of the local anopheline species responsible for malaria transmission and their breeding and feeding habits. In some cases draining swamps through the construction of ditches dried out the land and eliminated one set of vectors; but if the drainage ditches were not maintained, they became blocked, allowing vegetation to grow up on their banks. The ditches then provided sites for new malaria vectors. Multiple efforts to colonize the Roman Campagna and Pontine Marshes through drainage projects failed, as the areas remained malarial.[5]

MEDICAL SCIENCE AND THE NEW PUBLIC HEALTH

The development of more effective tools for preventing malaria and reducing its effects on human health awaited the discovery of malaria parasites and the role of anopheline mosquitoes in their transmission at the end of the nineteenth century. Alphonse Laveran's discovery of the malaria parasite in human blood in 1880; Ronald Ross's demonstration of the role of the anopheline mosquito in the transmission of malaria in birds, published in 1898; and Giovanni Grassi's conclusion in the same year that anopheline mosquitoes also transmitted malaria in humans provided the scientific basis for the development of a series of new approaches for combating malaria as well as for improving existing methods of control. Others have detailed the histories of these discoveries and the quarrels and debates they generated.[6] What interests us here is how these discoveries, together with subsequent research on the life cycles of malaria parasites and the breeding and feeding habits of the various anopheline mosquitoes capable of transmitting malaria, created a new model for understanding this ancient disease—a model that largely excluded from vision the broader social and economic contexts within which malaria occurred.

The new malaria model was part of the broader revolution in medicine and public health that occurred in the second half of the nineteenth century. Central to this revolution were the bacteriological discoveries of Louis Pasteur, Robert Koch, and Joseph Lister. Drawing on earlier developments in chemistry, physiology, pathological anatomy, and clinical medicine and the invention of the achromatic microscope—which permitted scientists to view bacteria—bacteriologists in the 1880s identified the organisms responsible for most of the major killer diseases that affected humans.

These discoveries did not transform medicine or public health practice overnight and, with few exceptions, did not lead immediately to new ways of treating disease. Earlier notions of disease causation, including ideas about environmental miasmas and individual constitutional predispositions to disease, continued to persist. Many physicians were openly critical of germ theory and its value for understanding and combating disease. But, even so, the bacterial revolution helped transform public health practice in fundamental ways and encouraged the adoption of a new public health model, which narrowed the focus of disease control efforts.

Public health practice in Europe and America during the first half of the nineteenth century had focused on social reform and the elimination of a broad range of conditions that were viewed as causing disease. Much of the work was aimed at improving the environment; eliminating waste and other sources of pollution; and improving the water supply, housing, nutrition, and conditions of employment. Disease could be prevented by improving the overall health and well-being of society. As historian Elizabeth Fee has noted, this vision of public health "required a diverse set of disciplines and skills: economics, sociology, psychology, politics, law, statistics and engineering, as will as the biological and clinical sciences."[7]

In the wake of the bacterial discoveries of the 1880s, however, a new vision of public health emerged, aimed at ensuring health by identifying and attacking the germs that caused disease. The bacteriological laboratory became the central symbol and weapon in this new pursuit of health. In place of the diverse collection of professionals who had driven health reform earlier in the century, bacteriologists and physicians peopled the new public health. In place of searching for the causes of ill health in society as a whole, the new public health professionals sought them under the microscope. Powerful new methods of identifying disease overshadowed efforts to improve water supplies, clean streets, and improve the housing and general living conditions of the poor. The declining importance of wider social and economic conditions was perhaps captured best by Pasteur himself when he stated, "Whatever the poverty, never will it breed disease"[8]

This view of scientific hygiene developed in Germany at the end of the nineteenth century and was subsequently transported to the United States, providing a model for the Johns Hopkins School of Hygiene and other schools of public health. It also permeated the work of the new international health organizations that arose at the end of the First World War. While the postwar period saw the rise of social medicine with its commitments to social welfare and the link between medicine and underlying social causes of disease, both the League of Nations Health Office and the International Labor Organization translated these concerns into the universal language of science. This led, during the 1920s, to a series of narrowly defined technical responses to health problems.

The International Labor Organization was premised on the assumption that social and economic conditions determined workers' welfare, yet its work was dominated by physiological studies carried out in laboratories and by the production of industrial standards. The League of Nations Health Office was concerned with the standardization of mortality statistics and providing a quantitative basis for chemotherapeutic drugs and vaccines. The league's Tuberculosis Committee focused on the technical issues related to the use of the French bacille-Calmette-Guérin (BCG) vaccine to prevent tuberculosis infection. Social deprivation, diet, and overall health conditions of the population were not considered.[9]

The discoveries of Laveran, Ross, and Grassi occurred at a time when the practice of public health as a whole was becoming increasingly focused on the identification and elimination of microparasites. It is thus not surprising that these discoveries led to a similar concentration of malaria control efforts on eliminating the proximate cause of the disease—parasites and mosquitoes—and a neglect of the broader social and economic conditions that played an important role in the epidemiology of the disease. Many of those who advocated the adoption of these new approaches recognized the important role that broader societal forces played in the epidemiology of malaria. Nevertheless, early efforts to translate biological knowledge into malaria control led public health officials to adopt policies that focused on attacking the newly discovered sources of the disease.

Investigation of the biology of malaria transmissions did not provide a simple answer to malaria control. To the contrary, it led to competing schools of thought with regard to how this knowledge could best be translated into preventive action. The life cycle of malaria parasites provided, and continues to provide, multiple points of attack. In general, however, public health officials at the beginning of the twentieth century were divided between those who advocated attacking the mosquitoes responsible for transmitting malaria among human hosts and those who called for an attack on the malaria parasite within the human host—to "treat the patient, not the mosquito," as Robert Koch argued.[10] Both approaches had advantages and disadvantages.

Vector Control

Vector control programs during the 1910s and 1920s focused on eliminating breeding sites through drainage and attacking larvae by spreading oil on the water surfaces on which female *Anopheles* laid their eggs. Paris Green proved to be particularly effective as a larvicide. In addition, some programs attacked adult mosquitoes with pyrethrum sprays. Attacking mosquitoes and their larvae had two advantages. First, it required little cooperation from the population at risk. Instead, small teams of malaria workers could be recruited to perform the necessary environmental changes. This was a particularly attractive feature for health officials who often held a low opinion of the ability of populations at risk to observe sanitation regulations and protect their own health. These attitudes were particularly prevalent among Europeans working in tropical colonial settings, as well as among health workers in the American South. A second advantage of vector control was that environmental engineering could contribute to a long-term reduction, or even elimination, of malaria.

Vector control, however, had two disadvantages. First, it required the leadership of trained entomologists who could identify the relevant anopheline vectors and their breeding habits and develop appropriate strategies for eliminating them. Earlier efforts to eliminate malaria through drainage without such knowledge had met with mixed results and in some cases made matters worse. The failure of the early British experiment in vector control at the military encampment at Mian Mir in India from 1902 to 1909 was due in part to a lack of understanding of the precise breeding conditions of local malaria vectors as well as to poor planning. It was also hampered by a lack of funds, which leads us to the second disadvantage of vector control methods: they were expensive to conduct.

While the cost of the Mian Mir experiment was never definitively calculated, one estimate put the five-year cost at 70,000 rupees. This would be the equivalent of nearly a half a million dollars today. Clearly, this was an amount which, if multiplied by the number of areas in need of malaria control in India, would have been enormous. Realization of potential costs of vector control played a significant role in the colonial Indian government's decision not to extend vector control more widely, even though those involved in malaria control gave considerable lip service to its advantages.[11]

The cost of vector control measures limited their use primarily to towns and cities and to sites of economic production, such as plantations, railway works, and mines. Nowhere was vector control employed broadly to protect large areas of rural settlement.

Yet even within this limited context, vector control, when well planned and executed, had dramatic results. Most notable was the work of Malcolm Watson in controlling malaria on rubber and tea plantations in Malaya in the early 1900s. Watson concluded that the Chinese labor force would not be receptive to using quinine and that the bed nets, screening, and oiling would be too costly. He decided instead to drain the marshes and fill in all the small bodies of water. Watson's methods were adopted and modified by Dutch entomologist N. R. Swellengrebel in Indonesia in 1913. Instead of wholesale drainage, Swellengrebel identified the specific species of *Anopheles* that was responsible for malaria transmission locally and then eliminated the particular breeding sites they employed. This method became known as "species sanitation."

Perhaps the most famous application of vector control was

that instituted by Colonel William Gorgas to eliminate malaria and yellow fever in Panama. The two diseases had wreaked havoc on workers and contributed to the failure of French efforts to build a canal across the Isthmus of Panama at the end of the nineteenth century. By the time the French called it quits in 1888, some 20,000 workers had died, the vast majority from yellow fever and malaria. The Americans, who took over the project in 1902, fared no better during the first years of operation. As a result, the original engineer was replaced by John Stevens, who had built the Great Northern Railroad across the Pacific Northwest. Stevens concluded that the canal could not be completed without a well-fed, well-housed, and disease-free labor force. So, before beginning to dig, he recruited a group of military sanitation experts led by Colonel Gorgas, who had played an important role in the earlier elimination of yellow fever and malaria in Havana, to undertake the necessary sanitary measures. Gorgas began operations in 1904, and yellow fever was gone within two years. Mosquitoes of the genus *Stegomyia* were transmitting yellow fever in Panama, and their limited flight range made it relatively easy to eliminate adults and larvae by fumigating houses and emptying, covering, or oiling all water receptacles. In addition, workers who contracted yellow fever and survived became immune to the disease. Thus the human reservoir of infection declined over time. Malaria took longer to be brought under control because of the much larger area in which local vectors could breed and their greater range of flight. Malaria was also more difficult because workers could be repeatedly infected with the disease, and so the reservoir of infection remained large as long as transmission continued.

To combat malaria, Gorgas divided the canal zone into 25 districts.[12] Each was placed under a sanitary inspector who oversaw from 20 to 100 laborers. Having identified the breeding habits of the local malaria-carrying mosquitoes, each team set out to eliminate potential breeding sites. The workers cleared brush and undergrowth within 200 yards of houses and villages located within 5 miles of the canal. They drained all marshy areas, filling in drainage ditches with stone, or lining them with cement, to prevent the growth of vegetation that would have encouraged breeding. Gorgas calculated that his men drained 100 square miles of territory, "constructing in all some five million feet of open ditch, some one and a half million feet of concrete ditch and some one million feet of rock-filled ditch." Where it was not possible to drain an area because of its size, or where new breeding sites were created by construction activities, Gorgas directed his army of sanitary workers to apply kerosene oil. He estimated that they applied about 50,000 gallons a month over the 100 square miles of territory being treated. The inspectors also put screens on the doors and windows of the several thousand buildings, including worker barracks, that the Canal Commission constructed. Finally, while vector control was the primary method employed to combat malaria, prophylactic quinine was also used. Quinine was provided freely to all workers along the construction line at 21 dispensaries. In addition, quinine dispensers were on all hotel and mess tables. On average, half of the work force took a prophylactic dose of quinine each day.

Gorgas's sanitary activities greatly reduced the burden of malaria in the Canal Zone. The death rate due to malaria in employees dropped from 11.59 per 1,000 in November 1906 to 1.23 per 1,000 in December 1909. Deaths from malaria in the total population declined from a maximum of 16.21 per 1,000 in July 1906 to 2.58 per 1,000 in December 1909. Among the work force, the percentage of employees hospitalized due to malaria was 9.6 percent in December 1905, 5.7 percent in 1906, 1.8 percent in 1907, 3.0 percent in 1908, and 1.6 percent in 1909. Malcolm Watson called the sanitation of Panama, "the greatest sanitary achievement the world had seen."[13]

Attacking the Parasite

Attacking malaria parasites also had its advantages and disadvantages. The primary advantage was that it was generally less expensive to protect local populations with quinine than with the application of vector control methods. This permitted public health officials to achieve greater geographical coverage. The availability and price of quinine fluctuated, however, and during certain periods, when access to supplies were limited, rising prices could make the use of quinine extraordinarily expensive (if supplies could be had at all). But in general, attacking the malaria parasite was less expensive than eliminating its breeding sites. The second advantage of focusing on the parasite was that it did not require research or specialized knowledge of the malaria vectors. All that was needed was an effective drug distribution system.

The primary disadvantage of attacking the parasite was that it required the cooperation of the populations at risk. People had to be educated about the benefits of taking quinine at regular intervals during the transmission season. This process was complicated by quinine's extremely bitter taste. The second disadvantage of this approach was that it did not significantly affect the transmission of malaria or lead to any long-term decline in the incidence of the disease. Quinine had to be distributed and taken indefinitely, or until other measures were instituted that reduced transmission levels.

The most successful effort to control malaria through the distribution of quinine occurred in Italy between 1901 and 1910. As we saw in chapter 3, Italian health officials at the end of the nineteenth century faced a massive problem of malaria morbidity and mortality, particularly in the southern half of the country, where malaria was present everywhere, except in the central mountain range. Not only a health problem, the disease also seriously handicapped the economic development of the newly unified Italian state.

Soon after unification in 1870, the Liberal government attempted to control malaria in the Pontine Marshes outside of Rome by draining the swamps. As noted earlier, the project, instituted without knowledge of the role and breeding habits of local anopheline mosquitoes, failed to achieve its goal, and malaria remained a problem. The government rededicated itself to eliminating malaria following the scientific discoveries of the 1890s, a number of which had been made by Italian malariologists. Vector control methods were considered, but they were expensive and difficult to apply over all the infected areas of the country. Quinine, by comparison, could potentially reach everyone if its cost could be reduced. To achieve this goal, the government established a monopoly over the marketing of quinine and subsidized the cost of the drug by charging a quinine tax to landlords and other employers of outside labor, whom they viewed as responsible for the persistence of malaria. The government also made the drug available to those who could not afford even the low subsidized price, through a network of some 1,200 rural health centers constructed by the beginning of World War I. Finally the government initiated an extensive public health campaign linked to the creation of peasant schools. These schools provided general education and not just malaria information. They reflected the belief that general social uplift could also contribute to the disappearance of malaria.

Although the campaign did not achieve all its goals, it was highly successful in one respect. It contributed to a major reduction in malaria mortality in Italy, which declined from 490 per million in 1900 to 57 per million in 1914. Yet morbidity rates over the course of this period, while fluctuating from year to year, remained essentially constant. Despite the hopes of Grassi and Celli that the massive distribution of quinine could lead to the eradication of malaria in Italy, the quinine campaign had little impact on transmission. The reason is that quinine attacked only the merozoite stage of the malaria parasite. This reduced the symptoms of the disease and limited mortality, but it did not prevent humans from being infected, nor did it eliminate the gametocytes that infect mosquitoes. Thus transmission continued. In addition, despite their massive public health campaign, Italian authorities found it difficult to convince otherwise healthy individuals to take a bitter tasting medicine to prevent disease. A large amount of the quinine distributed to local dispensaries was never used. The failure to eliminate or significantly reduce transmission had important consequences: when the quinine distribution system broke down during World War I and wartime conditions led to increases in transmission, malaria mortality rose dramatically.[14]

Supporters of each approach to malaria control were widely distributed within Western medical circles.[15] Many believed that in combating malaria health officials should use whatever method was available. But from the 1890s through the 1920s, opinion was broadly divided between public health leaders in America (and, to a lesser extent, Great Britain) and those in most of continental Europe. The former tended to favor attacking the mosquito, whereas the latter favored attacking the parasite and protecting human hosts, primarily by using quinine. The support of Ameri-

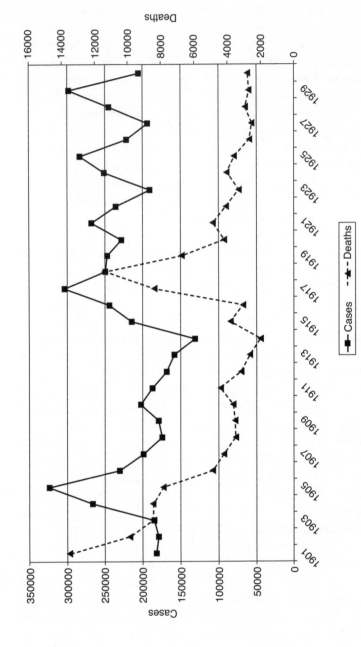

Malaria Cases and Deaths in Italy, 1900–1930

Source: J. A. Najera et al., *Malaria Epidemics, Detection and Control Forecasting and Prevention*, Geneva, 1998 (WHO/MAL/98. 1084), 15.

can health officials for vector control stemmed in large measure from their early successes in controlling yellow fever and malaria in Cuba and Panama. Quinine had also proved unsuccessful in controlling malaria in the American South. British medical authorities played a significant role in developing vector control methods, following the work of Ross and Malcolm Watson. Yet British medical opinion was divided over whether vector control methods were economically feasible. Patrick Manson, the father of British tropical medicine and an early supporter of Ross's work, felt that active sanitation programs to combat tropical diseases in the colonies were expensive and politically impractical. Nonetheless, during the First World War, vector control measures were used with some effect to protect British and Indian troops fighting in malarious regions. Quinine prophylaxis, on the other hand, proved to be less useful.[16]

On the other side of the divide, the preference of health officials in continental Europe for attacking the malaria parasite was greatly influenced by the dominance of bacteriology on the continent. France's Louis Pasteur and Germany's Robert Koch, the fathers of bacteriology, were convinced that malaria was no different than other bacteriological infections, in that it could be best treated and prevented by attacking the germ that caused the disease, the malaria parasite. Two additional factors led European health officials to downplay the usefulness of vector control. The first was differences in the dominant malaria vectors in Europe and America. In America, the most prevalent mosquito vector, as we saw in chapter 2, was *A. quadrimaculatus.* This mosquito bred among the vegetation in still bodies of water. As such, it was particularly susceptible to larvicides. The primary malaria vector in Europe was *A. maculipennis.* This mosquito bred on the edges of clean flowing water. It was much less susceptible to larvicides until the development of Paris Green in the 1920s.

A more important problem for European observers was the existence of numerous areas in Europe where *A. maculipennis* appeared to breed in abundance, and yet where there was no malaria. This phenomenon, referred to as "anophelism without malaria," caused many to question the value of applying vector control methods.

Although malaria control strategies from the end of the nineteenth century followed two distinct pathways, they were similar in one important respect. They were both rooted in a biological construct of disease that largely neglected the underlying social and economic conditions, which, as we have seen in earlier chapters, often drove the transmission of malaria. This did not mean that the advocates of vector control or the use of quinine to battle malaria were blind to the economic and social conditions that were often associated with intensive malaria transmission; or that they failed to acknowledge the role which improvements in these conditions played in the elimination of malaria. But, up through the early 1920s, these observations played only a limited role in the development of malaria control efforts. Thus, while the Italian malariologist Angelo Celli pointed to the exploitation of workers as a major reason for Italy's malaria problem in his malaria textbook, he contented himself "with comparing two columns of figures, one showing increasing sales of state quinine and the other falling rates of death," while running the state's antimalaria program.[17] Similarly, M. A. Barber recognized the importance of social and economic advancement in the disappearance of malaria but focused his own work in the American South on vector control activities.

It was not until later in the 1920s and 1930s that antimalarial programs incorporated general improvements in economic and social welfare, and then only briefly. The move toward these broader approaches was signaled by two reports of the League of Nations Malaria Commission, the first of which appeared in 1924.

Malaria Control and Rural Development

The League of Nations was founded, after the Paris Peace Conference of 1919, to prevent future wars by settling disputes between member countries and by improving global welfare. One of its commissions was the League of Nations Health Organization, which created a subcommission, the Malaria Commission, to deal with wartime upsurges in malaria in many European countries. The Malaria Commission's members were asked to examine malaria control measures in various member countries and to make recommendations concerning the most effective strategy for combating the disease. But the real purpose of the commission was to resolve the division between supporters of vector control and those who preferred to focus on the human host and the malaria parasite. The commission's first report, issued in 1924, placed great emphasis on the need to protect human populations through a combination of "quininization" and improvements in housing, education, nutrition, and agriculture. It concluded:

> The Commission feels bound to reiterate the importance of the general social and hygienic condition of the people in relation to the extent and severity with which malaria shows itself. This has been brought to notice in many areas; better housing, an ampler and more varied dietary, and better environmental conditions make for a more intelligent and willing people and for greater individual resistance. Moreover, in such conditions, quinine is utilized much more intensively and effectively. The improved economic conditions of a country not only induce optimism and tend to stimulate the inertia of misery into organized activity but allow for the extension of health services and the co-operation, in measures for their own good, of the people whose health the sanitary service is designed to improve.[18]

The report was less supportive of vector control measures, which were treated as a kind of second line of defense. Two of the commission's members visited the United States in 1927. In their report of the visit, they chose to largely ignore the vector control activities being carried out by American health authorities, emphasizing instead the role of social and economic progress in the decline of the disease. The commission's second general report, *Principles and Measures of Antimalarial Measures in Europe,* reiterated the commission's support for quinine and its earlier emphasis on the importance of general social uplift. However, it was much more critical of vector control measures, claiming, "they led to exaggerated expectations, followed sooner or later by disappointment and abandonment of work." The report went on to note, "In some countries . . . hardly anything has retarded the effective control of malaria so much as the belief that, because mosquitoes carry malaria, their elimination should be the object of chief concern and expenditure."[19]

The report's first author was British malariologist Colonel S. P. James. James's lack of enthusiasm for vector control was deep-seated and stemmed from his early experience with the failed Mian Mir experiment in India. It was also based on his experience in England, where, as we have seen, malaria largely disappeared even before the initiation of any antimalarial activities. James was strongly committed to the idea that the elimination of malaria followed from broader social and economic development. Visiting Kenya in the late 1920s, James recommended that the best way to eliminate malaria in the so-called Native Reserves, in which the majority of the colony's African population lived, was "to introduce agricultural, and in some cases industrial, welfare schemes which aim at improving the economic and social conditions of the people and their general well-being and standard of living."[20]

The Malaria Commission's appreciation of the role of social and economic advancement was influenced in part by early Italian success in eliminating malaria through a policy of generalized rural uplift known as *bonifica integrale,* or bonification, begun after World War I. Bonification sought both to eliminate malaria and to create new agricultural lands that would support poor farmers, allowing them to improve their economic status along with their overall health and well-being. The results of these enterprises were mixed. As we have seen, efforts to transform the Roman Campagna at the end of the nineteenth century failed. Yet there were some notable successes. According the League of Nations Malaria Commission, the Italian state invested 40 million lire in bonification projects between 1919 and 1925, and private consortia contributed another 25 million. Just over 700,000 hectares were transformed, most in the northern part of the country. The populations of the improved areas grew from 441,340 to 724,073, and the value of agricultural production grew from 137,918,800 lire to 1,048,527,677 lire, with the number of animals raised on the land growing from 103,439 to 241,358. Finally, malaria mortality dropped dramatically in these areas.[21] Even so, it is impossible to separate out how much of this reduction in malaria mortality was due to the improved conditions created by bonification and how much to quinine distribution, which continued during this period. Nevertheless, there was widespread belief that bonification had improved the malaria situation.

Bonification achieved its most spectacular success under the Fascist government of Mussolini, which came to power in 1922. Mussolini's government, like its predecessors, saw the persistence of malaria as a challenge to its legitimacy and capabilities. It also saw bonification as a means of building political support, particularly in the northern and central areas of Italy. Mussolini hoped to create a new class of yeoman farmers, who would support the state and serve as a buffer against the populist movement that had grown up in the South in association with the earlier quinine campaign and its antilandlord ideology. Of particular symbolic importance was the elimination of malaria in the Roman Campagna and Pontine Marshes, a goal that had eluded previous governments, as well as a long line of pontiffs and princes.[22] The marshes had a total population of 1,637 people in 1928 with a population density of 7 people per square kilometer. Only 1 in 7 of these were engaged in agriculture. The Red Cross estimated that 80 percent of those who spent a single night in the marshes during the malaria season contracted the disease.[23] By reclaiming the land, and eliminating malaria, Mussolini's government hoped to attract a new population of farmers to the area.

Fascist bonification efforts began at Maccarese in the Tiber Delta in 1926. In 1927 the director of public health drew up plans for eliminating malaria in the Pontine Marshes, and work began in 1929. Italian engineers began by constructing a massive dam at the foot of the Lapini Mountains to collect mountain water that fed the marshes. The water was redirected to a central canal, named after Mussolini, which ran 40 kilometers to the Tyrrhenian Sea. Once the dam and canal were built, excess water was drained from the surrounding swamps through a network of drainage ditches, which led to collection areas or sumps. Because much of the marshes were located below sea level, the collected water had to be raised to the level of the central canal by means of a series of hydraulic pumps.

The reclaimed land was then cleared of undergrowth. New roads were constructed and new agricultural colonies were

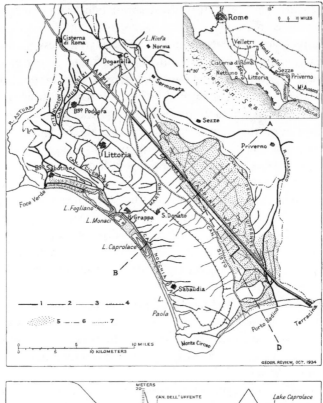

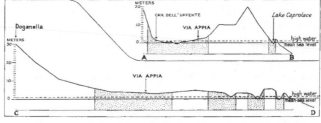

Reclamation Plan for the Pontine Marshes, 1934

Source: Ruth Sterling Frost, "The Reclamation of the Pontine Marshes," *Geographical Review* 24, 4 (1934): 585.

founded, each with its church, hospital, and social club. In addition, the state established rural schools and health centers. At Quadrato, an industrial center was constructed to manufacture windows, doors, and furniture for the settlers' new houses. The center also repaired the machinery used for the bonification project. Experimental gardens were created under government control to determine which crops were best suited for the different parts of the reclaimed countryside. All of these measures were included under the heading of "Grand Bonification," for which the government paid 75 percent of the cost, the rest being borne by the local authorities and the owners of the land.

The bonification project also incorporated efforts directly aimed at eliminating malaria transmission. These included the screening in of settler houses and the use of gambusia fish and Paris Green to eliminate larvae. In Maccarese, planners also placed cattle in close proximity to human habitations to attract local anopheline vectors. Cattle stalls were brought under the same roof as the dwellings, after the latter were protected with wire gauze, and the cattle were not allowed in the fields. Finally, inhabitants were encouraged to take quinine to prevent malaria. A government quinine factory at Turin turned out 6,000 kilograms annually.[24]

By 1939 the population of the "bonified" areas of the Pontine Marshes reached 60,000 and 200,000 acres of farmland had been reclaimed. Where the malarious swamps had for centuries inhibited human settlement, a thriving new province had been created. Mussolini called it Littoria after the symbol of Fascist rule. The reclaimed marshes were widely proclaimed as a victory for Mussolini's Fascist state.[25]

The success did not come without costs. The direct financial costs of bonification were roughly 549 million lire, which is equivalent to roughly US$400 million in 2005 dollars. The human costs were less easily calculated. To begin with, the original inhabitants of the marshes, poor and debilitated by recurrent malaria infections, were displaced by the project, a pattern that would be repeated in more recent efforts at malaria control in other parts of the globe. In addition, little consideration was given to the health of the thousands of workers who were recruited to drain the swamps and construct fields and housing for future settlers.

Nonetheless, the Italian bonification projects provided a model for similar efforts elsewhere in southern Europe, though nowhere were they carried out as extensively as in Italy. The Italian experience also provided a model for malaria control efforts outside of Europe. In Palestine during the 1920s and 1930s, the Malaria Research Unit under the British Mandate government initiated a program of rural development in areas in which malaria was a significant problem. It employed vector control methods to attack anopheline-breeding sites beginning in 1923. These methods were combined in 1925 with efforts to improve agricultural practices, enlarge the amount of land under cultivation, improve water supplies and sanitary facilities, raise the number of schools and availability of education, and increase access to general health services. Malaria prevalence, as measured by the percentage of rural inhabitants with enlarged spleens, a symptom of chronic malarial infection, declined from 12 percent in 1925 to 5.4 percent in 1940. Improvement in overall health was reflected in a reduction of the infant mortality rate from 163 per 1,000 in 1926 to 76 per 1,000 in 1946.[26]

One of the most ambitious programs of integrated rural development linked to malaria control was initiated in the United States under the Tennessee Valley Authority in the early 1930s. In 1933, 3 million people lived in the area that was to become part of the TVA project. Their average income was 45 percent of the national average. The population was made up primarily of tenant farmers, and unemployment was high. Much of the land had been farmed too hard for too long, which eroded and depleted the soil. Crop yields had fallen along with farm incomes. The best timber had been cut. Soil quality was poor and malaria was endemic. Existing river and stream impoundment throughout the region had provided ideal breeding sites for *A. quadrimaculatus.*

The TVA was intended to develop an integrated approach to rural development in which every aspect of development would be coordinated. The TVA developed fertilizers, taught farmers how to improve crop yields, and helped replant forests, control forest fires, and improve habitat for wildlife and fish. The most dramatic change in valley life came from the electricity generated by TVA dams. Electric lights and modern appliances made life easier and farms more productive—and electricity also drew industries into the region, providing desperately needed jobs.

The dams also created reservoirs encompassing more than 11,000 miles of shoreline. The creation of this shoreline threatened to increase the problem of malaria, by providing additional breeding sites for *A. quadrimaculatus*. To prevent this from happening, the TVA employed careful environmental management methods based on knowledge of the breeding habits of *A. quadrimaculatus* mosquitoes. These included changing of the water level of the reservoirs periodically, through the release of water. *A. quadrimaculatus* laid its eggs along the shoreline, so lowering the water led to their desiccation. In addition, TVA officials conducted the annual removal of vegetation that might be conducive to mosquito breeding. The goal of these measures was to prevent the population of adult vectors from reaching the threshold level needed to promote transmission. The TVA supplemented these environmental management techniques with additional control measures, including house-proofing with screens and the use of larvicides, especially on smaller impounded areas of water found throughout the TVA region. The TVA also encouraged the creation of cattle grazing areas in close proximity to shoreline areas. This served several purposes. It helped eliminate plants associated with anopheline breeding, reduced the biomass of troublesome semiaquatic mosquito species, and limited the movement of *A. quadrimaculatus* away from the shoreline breeding sites, because the mosquito could feed off the cattle. Finally, cattle raising had additional economic benefits for local farmers. As a result of these integrated measures, malaria dropped dramatically between 1933 and 1938. By the end of World War II, when DDT was introduced, malaria had all but disappeared from the TVA region.[27]

The economic benefits of the TVA are more difficult to measure. At best they were distributed unevenly. Many families were dispossessed from their lands by the creation of dams. For the region as a whole, however, there is little doubt that the TVA brought improvements in the standard of living and many new economic opportunities.[28]

Not all efforts at bonification were successful. In northern Argentina in the 1930s large drainage projects successfully expanded agricultural production but did not result in a reduction of malaria. The preferred breeding places of the primary malaria vector in the region, *A. pseudopunctipennis,* were not the marshes and bogs that were eliminated, but brooks and rivers that had sunny banks and in which *Spirogyra* algae were abundant. Unfortunately for the Argentinean malaria controllers, *Spirogyra* also grew in the stone or concrete faced canals built to drain the land. Thus the bonification efforts actually increased the number of breeding places in the areas being reclaimed.[29]

The new emphasis on more broad-based integrated approaches to malaria control in the late 1920s and early 1930s reflected a broader shift in public health thinking during this period. This change in direction was driven by the work of several international health organizations.

The League of Nations Health Organization (LNHO), under the leadership of the Polish-born public health leader Ludwig Rajchman, began developing a program for rural social medicine in the late twenties. The new program drew on eastern European models and reflected a more integrated view of health and development in which health was linked to advances in general social and economic welfare. This shift was reinforced by the global Depression that revealed in stark clarity the importance of economic status for health.

The LNHO joined with the International Labor Organization in a series of studies and conferences on how diet, housing, and economic conditions affected health. The two organizations also carried out surveys of rural hygiene and analyzed the relationship between public health and sickness insurance. In 1932 the International Labor Organization issued a report on the "economic depression and public health." The report called for studies of morbidity, nutrition, and the psychological effects of unemployment, as well as the effects of poverty on children.

The LNHO's new emphasis was also reflected in the work of its Tuberculosis Commission. In the 1930s the commission moved from measuring vaccine efficacy to examining the importance of wages, working hours, diet, and living conditions in the epidemiology of the disease. The new emphasis was also reflected in the LNHO's international activities.[30]

Although the league's Health Organization was primarily concerned with conditions in Europe, Rajchman believed strongly that the LNHO should play a role in helping poor countries everywhere. However, many potential areas of development fell under the colonial domination of member states and were viewed as politically out of bounds. Because China did not fall under the direct colonial domination of a member state, it provided great opportunities for making improvements in health. Rajchman enlisted two fellow east Europeans, Croatians Berislav Borčič and Andrija Štampar, who had both been on the faculty of the Zagreb School of Public Health, to develop a Rural Rehabilitation Program in the Kaingsi Province of China. The program, begun in 1933, combined social and economic development efforts with improvements in health services, including malaria control. Reflecting on his experience directing the program, Štampar wrote that "After working nearly three years in China, I am especially impressed by one fact. Successful health work is not possible where the standard of living falls below the level of tolerable existence. Public health policy must be intimately connected with a programme for general social improvement."[31]

The LHNO was not alone in developing integrated health and development programs in China. The Rockefeller Foundation's International Health Division embarked on an ambitious program in northern China in 1935.[32] The program brought together personnel from the Institute of Rural Administration (Yenching), the Mass Education Program in Ding County, the Nankai Institute of Economics, the University of Nanking's College of Agriculture, the Department of Preventive Medicine at the Peking Union Medical College, and the North China Industrial Institute. The North China Program represented a significant break from the foundation's traditional focus on more narrowly defined biomedical approaches to public health. It was directed by the foundation's Selskar Gunn, with the strong influence and support of John Grant, who was director of the Department of Preventive Medicine at Peking Union Medical College and a Rockefeller Foundation officer. The North China Program, which received a three-year commitment of funds from the foundation in the amount of $1 million, was more important for the vision of health and development it represented, than for its actual accomplishments. While the program would later serve as a model for the famous barefoot doctor program under the Communists, its ability to function was seriously hampered by the war that began with Japan's invasion of Manchuria in 1937. Gunn left in 1938, and although the program continued to function the following year, it did not survive the war.

In 1937 the LNHO organized an Intergovernmental Conference of Far-Eastern Countries on Rural Hygiene, held in Bandoeng, Indonesia. The conference dealt with a wide range of issues and blended discussions of focused public health programs with calls for broader social and economic development. The summary conference report, written by Selskar Gunn, concluded, "One thing is certain . . . unless the economic and cultural level of the rural populations can be raised, there can be no hope of employing curative or preventive measures with any degree of success."[33]

Thus a more expansive approach to malaria and development emerged in the 1920s and 1930s as part of a broader movement in public health away from narrow biomedical approaches and toward a more integrated vision in which health was seen as intimately linked to social and economic development. But this vision did not survive the interwar period. By the end of World War II, a more narrow vision of health based on faith in science and the use of new biomedical technologies again dominated the field of public health.

THE GROWING HEGEMONY OF VECTOR CONTROL

Supporters of vector control had never given up faith in the value of their approach. Vector control methods, to subdue malaria, had continued to be used by public health authorities throughout the 1920s and 1930s. As we have seen in the cases of Italian bonification, the TVA, and Palestine, vector control was an integral part of programs of social and economic uplift. What was contested during the late 1920s and early 1930s was the extent to which vector control by itself was sufficiently effective to dramatically reduce or eliminate malaria transmission. Opponents believed strongly that it was not. By the end of World War II, however, few people questioned that it was, and vector control became entrenched as the single most important means to attack malaria.

Unraveling Anophelism without Malaria

Four developments moved vector control strategies to this dominant position. The first involved advances in knowledge about anopheline mosquitoes. Since the beginning of the twentieth century, opponents of vector control had pointed to places where anopheline mosquitoes capable of transmitting malaria existed, but where there was little or no malaria. Why, they asked, should large amounts of money be spent on eliminating malaria vectors, when their presence did not necessarily cause malaria? Was it not cheaper and more efficient to treat cases with quinine when they occurred or to use quinine prophylactically? The problem of "anophelism without malaria" had long puzzled entomologists. Numerous studies had been carried out to determine why known transmitters of malaria did not transmit the disease. Perhaps, some reasoned, malaria depended on the susceptibility of the population at risk. Predisposition to infection remained a major tenet of Western medicine. If so, did it not make sense to improve the general health and well-being of populations at risk? Critics also questioned why malaria had disappeared from large areas of Europe, without any effort to eliminate mosquitoes. Again, this seemed to point to the value of wider improvements in health.

Research contributed to a more detailed knowledge of the morphology and behavioral characteristics of various species of anopheline mosquitoes. Yet it was not until the early 1930s that this research began to provide answers to the problem of "anophelism without malaria." Rockefeller officer Lewis Hackett joined forces with Italian malariologist Alberto Missiroli to unravel the puzzle. What they discovered was that the primary malarial vector over much of Europe, A. maculipennis, was not a single species, but in fact six separate species. While the six looked very similar, they exhibited different feeding and breeding behaviors. Some preferred human blood, others fed off of animals.[34] This explained why large populations of A. maculipennis could exist in an area without malaria transmission. It also helped explain why malaria could disappear from certain regions without any effort to eliminate mosquitoes. Ecological changes brought about by social and economic development could lead to a change in the distribution of maculipennis species, in which human-feeding Anopheles were replaced by a species that preferred animal blood. For example, as we saw in chapter 2, in many parts of western Europe where the intensification of agricultural production eliminated grazing lands and led to an increase in the housing of cattle in close proximity to human habitation, species of *Anopheles* that took their blood meals from both animals and humans squeezed out a species that fed predominantly on humans. This led to a reduction in human transmission. Research into the problem of "anophelism without malaria" provided a more precise understanding of the conditions under which various anopheline species transmitted malaria. In doing so, it helped eliminate much of the mystery that had clouded understanding of the role of anopheline mosquitoes in malaria transmission. Henceforth, it could be argued that the key to the distribution of malaria was the presence or absence of man-biting *Anopheles* in close proximity to human populations. Where man-biters did not exist, or were dominated by other species as a result of ecological changes, either malaria did not occur or it disappeared.

Fred Soper and Species Eradication

The second development that solidified the value of vector control approaches to malaria, was the victory of Frederick L. Soper of the Rockefeller Foundation's International Health Division in eradicating the *Anopheles gambiae* mosquito from Brazil between 1939 and 1941. As we saw in chapter 4, this mosquito was accidentally transported by ship to northeastern Brazil from West Africa in 1931. The mosquito quickly established itself and contributed to a series of malaria epidemics during the 1930s. Soper, who at the time was heading a yellow fever eradication effort in Brazil, requested funds from the foundation to eradicate the gambiae mosquito. Employing a disciplined army of workers to attack the mosquitoes' breeding sites and to fumigate houses, cars, and trucks, Soper eliminated the invader in two years, greatly reducing the prevalence of malaria sickness and death. Soper's victory was proclaimed in heroic terms by the Rockefeller Foundation in its widely read annual reports and presented as vindication of vector control methods. Soper himself viewed his success as proof that vector eradication was a cost-effective means of eliminating malaria.[35] In a memo to the president of the Rockefeller Foundation in 1942, he concluded: "The eradication of *Anopheles gambiae* from a large known infested region of northeast Brazil is an important event in public health administration, not only because it relieves the continent of the immediate threat which existed, but because it dramatically calls attention to the possibilities of controlling mosquito-borne disease through species eradication rather than through species reduction."[36]

As we saw in chapter 4, the history of the 1938–39 epidemic in Brazil was not just about the invasion and subsequent eradication of the *Anopheles gambiae* mosquito. It was also about the changing social and economic conditions of the region. Moreover, improving economic conditions beginning in 1940 facilitated the success of Soper's campaign. Had the need for social and economic improvement been recognized, the history of the epidemic might have been used to argue for linking of vector control with broader efforts at social and economic reform. As it was, Soper failed to appreciate this wider context and, along with the powerful Rockefeller Foundation, employed the history of the epidemic to validate vector control methods for eliminating malaria. Soper went on to repeat his success in Brazil by eradicating *A. gambiae* mosquitoes that had invaded the lower Nile region of Egypt in the early 1940s. There gambiae mosquitoes had moved northward from the Sudan, producing major outbreaks of malaria in upper Egypt, and were threatening to move into lower Egypt where the British army was preparing to defend the Suez Canal from German advances. Soper once again organized a successful anti-gambiae campaign. This further reinforced his claims concerning the value of species eradication, though wartime secrecy prevented Soper or the foundation from singing the praises of Soper's most recent victory until after the war.[37]

By the early 1940s, the unraveling of the problem of "anophelism without malaria," together with Soper's victories in Brazil and Egypt, and a number of other successful uses of vector control to eliminate malaria in parts of the American South and Latin America, was beginning to shift the tide of public health opinion away from the position advocated by the League of Nations Malaria Commission and toward that advocated by supporters of vector control. This shift represented a movement from broader-based integrated programs to more narrowly defined technical approaches to malaria control. Two other developments accelerated this shift. The first was the outbreak of World War II. The second was the emergence of an increasingly conservative political climate in Europe and America, beginning in the late 1930s.

War and Pesticides

The outbreak of World War II brought to an end the Rockefeller Foundation's integrated experiment in North China. Yet it had more profound consequences for malaria control. Many of the theaters of operation during the war were located in regions where malaria was present, and both sides in the conflict suffered from the disease. Allied troops in the Pacific paid a heavy price until the middle of 1943, when more concerted efforts to develop a range of control methods, and their disciplined application, led to a gradual reduction in malaria cases.

Allied troops initially employed time-tested methods of vector control, using drainage, larvicides, and sanitation, to prevent malaria around military encampments. These methods were less effective, however, for protecting troops on the move. For this the military relied initially on personal protection through the use of long sleeved shirts and pants, mosquito repellants, and prophylactic drugs. The loss of Java to the Japanese during the early months of the war, however, greatly reduced the availability of quinine and forced medical researchers to look for synthetic alternatives. The most effective of these was atabrine, which was originally developed in Germany in the 1920s. The drug was sold to the Americans before the war and introduced to troops in the Pacific in 1942. The drug had side effects and caused the skin to turn yellow. In addition, it was unclear how much could safely be administered and for how long. The Australians, however, successfully developed a safe drug regimen, and by 1943 atabrine was being used widely among allied troops in Europe and the Pacific.[38] The U.S. antimalaria research program, much of it situated at the Johns Hopkins University, tested thousands of synthetic compounds and ratified the use of chloroquine by the end of the war.[39]

At the same time, efforts were made to develop new methods of vector control employing chemical pesticides. While the use of chemical compounds to control insects had attracted the attention of some entomologists prior to the war, this was but one of a number of fields of inquiry. Entomological research on malaria had focused primarily on the behavior of the subspecies of anopheline mosquitoes responsible for transmitting malaria in particular settings, studying the microenvironments in which *Anopheles* bred and lived, and developing ways of altering these environments in order to eliminate potential vectors.

The need to find simple yet effective methods to protect troops in tropical environments, to prevent disease outbreaks among displaced populations, and to increase food production at home led entomologists in a number of countries to concentrate their energies on pesticide development. This shift was particularly strong in the United States.

In December 1941, within weeks of the Japanese attack on Pearl Harbor, important groups of American entomologists called upon the U.S. Bureau of Entomology and Plant Quarantine to convene a meeting of entomologists to ascertain the best use of entomological resources for defense purposes. Over the next four years the entire entomological profession reoriented its activities to developing pesticides for wartime uses. Research policies that had favored long-term research agendas now encouraged projects designed to meet short-term military needs. Individual entomologists saw research on pesticides as contributing to the war effort.[40]

In mid-1942 American representatives of the Swiss firm J. R. Geigy Company brought to the U.S. Agriculture Department an insecticide that bore the trade name Gerasol. Tests over the next year showed that the compound, known as dichloro diphenyl trichloroethane, or DDT, had amazing killing powers. More importantly, once it was sprayed on the walls of human habitation it retained its toxicity for mosquitoes for months. This residual effect meant that spraying needed to be done only every six months, or in some cases once a year. This significantly reduced the cost of vector control. It also became clear that DDT's acute toxicity to humans was very low.

The discovery of DDT served as a catalyst for a profession that was already committed to developing pesticides. It resulted in a massive outpouring of money and a large investment of time by scientists and engineers to provide the full range of equipment for producing and using DDT. The U.S. Bureau of Entomology and Plant Quarantine developed spraying equipment, from aerosol bombs to tanks and nozzles for aerial delivery.[41]

By the end of the war the shift away from biological and cultural methods of control and toward chemical research was well established. The trend was reinforced by the flow of funds from chemical corporations into universities to support research into new insecticide compounds. The U.S. Department of Agriculture estimated that 25 new pesticides were introduced between 1945 and 1953.[42]

The pesticide industry grew dramatically, fueled by postwar demands for new chemicals to foster the growth of the world's food supply and protect populations living in developing countries from tropical diseases. According to one observer,

> The resultant expansion of the pesticide industry was so rapid that it simply steam-rollered [sic] pest-control technology. Entomologists and other pest control specialists were sucked into the vortex and for a couple of decades became so engrossed in developing, producing and assessing new pesticides that they forgot that pest control is essentially an ecological matter. Thus, virtually an entire generation of researchers and teachers came to equate pest management with chemical control.[43]

This description may overstate the shift that occurred in the field of entomology. However, there can be little doubt that by the time public health officials began to worry about their reliance on pesticides, entomology had little to provide in the way of alternative approaches.

If wartime requirements encouraged entomologists to focus their research on pesticides, abandoning research on other aspects of insect control, the success of DDT in controlling disease in Europe and Latin America both during and immediately after the war reinforced this trend. For these successes convinced experts in the newly constituted World Health Organization (WHO), as well as in Western agencies involved in international economic and social development, that research on malaria control, other than in the sphere of pesticide research, was unnecessary.

A number of experiments to test the effectiveness of DDT in

controlling malaria were launched immediately following the war. In 1944 and 1945 the Malaria Control Demonstration Unit of the Public Health Sub-Commission of the Allied Commission tested the use of DDT in house spraying to control malaria in the Pontine Marshes, where malaria had returned following the German destruction of the drainage system created by Mussolini.[44] These studies led to the conclusion that in areas where *A. labranchiae* was the only important vector, malaria could be readily controlled through a single application of DDT once a year. The Board of Scientific Directors of the International Health Division of the Rockefeller Foundation concluded in September 1947 that DDT had made earlier methods of malaria control superfluous in the Tiber Delta and Pontine Marshes. Drains, inverts, and the like were no longer important except for agricultural draining. Screening could also be forgotten if the DDT continued to be effective.[45] This conclusion represented a selective reading of the history of this experiment. As Frank Snowden has shown, DDT was not the only weapon used in this battle. Quinine was widely distributed. Paris Green was employed to kill larvae. Equally important, the inhabitants of the region had been exposed to decades of malaria control efforts. They were both informed about the disease and responsive to control efforts. Finally, Snowden notes that the campaign itself generated widespread employment that may have contributed to an overall improvement in health. But the Rockefeller Foundation chose to focus instead on the contributions of DDT.[46]

Two experimental programs in anopheline eradication were also begun in Sardinia and Cyprus in 1945–46 using DDT alone. The Rockefeller Foundation, the United Nations Relief and Rehabilitation Administration, and the U.S. Economic Cooperation Administration funded the Sardinia program, and the British Colonial Office funded the Cyprus experiment. While unsuccessful in eradicating malaria vectors, both programs succeeded in eradicating malaria, by reducing the numbers of anopheline mosquitoes below the threshold at which transmission could occur and treating all cases so as to eliminate the human reservoir of infection. Subsequently, large-scale control programs relying exclusively on house spraying with DDT began in Ceylon, India, and Venezuela in 1946.

Spectacular successes in each of these countries convinced the WHO Expert Committee on Malaria, which met in 1948, that DDT was an effective weapon for controlling malaria. "There is definite and overwhelming evidence that the recently introduced insecticides can be relied on as a basis for a widespread attack on malaria with the expectation of a significant reduction of morbidity in areas where they are properly applied."[47]

Dr. Paul Russell, who served on the Expert Committee, observed that, "We have tried very hard indeed to make the report a practical one and to emphasize primarily the usefulness of DDT residual spraying and the importance of getting this started in areas all over the world where nothing is being done"[48] Following the recommendations of the committee, WHO organized a series of demonstration teams to assist governments in 11 countries to develop malaria control programs based on the use of pesticides.

Notwithstanding these successes, subsequent expert committees supported the use of a range of control methods. The fifth expert committee, which met in 1953, noted, "Despite the great success of the newer forms of insecticidal work, there remains room for the use of more traditional methods. In some areas, constructional or drainage work alone may be indicated." The committee went on to warn that, "There is a risk that pride in the power [edited to read "preoccupation with the power"] of residual insecticides could result in derogation of other methods which have considerable remaining utility."[49] The committee's concern was well founded. Not only did the use of alternative methods decrease during the 1940s and early 1950s, but the growing preoccupation with DDT contributed to a decline in malaria research.

Malaria and Politics

Growing political conservatism in western Europe and America was the final factor that encouraged the movement away from broad-based integrated approaches to malaria control and toward the narrow application of pesticides. From the late 1930s through the 1950s, the political terrain upon which malaria control was debated in Europe and America shifted to the right. With this shift, an increasing number of European and American political leaders began to view models of social medicine that linked health to broader social and economic reform as being influenced by socialism and thus politically suspect. Evidence of this shift can be seen in the opposition of the U.S. surgeon general, Hugh Cummings, to LNHO director Ludwig Rajchman's proposal to hold a rural hygiene conference in Latin America, similar to the conference that took place in Bandoeng in 1937. Cummings objected to the proposal citing Rajchman's "social radicalism." More importantly, the shift was manifested in the purge of social medicine activists, including Rajchman, from the leadership of the League of Nations Health Organization in the late thirties. This shift to the right accelerated after the war and blocked any effort to seriously pursue broad, integrated approaches to malaria control in the postwar era.[50]

Prewar concerns about socialist influences evolved into postwar fears about communist expansionism. During the immediate postwar period, political leaders in western Europe and America became increasingly concerned about the growing power of the Soviet Union. The resulting Cold War shaped the development of malaria control programs in a number of ways. Its most direct impact was in encouraging the widespread use of DDT.

The apparent speed with which malaria could be brought under control with DDT, together with its short-term effects on other household pests, made malaria control particularly attractive for those who saw tropical disease control as an instrument for winning the support of local populations in the war against communist expansion.[51] The U.S. Special Technical and Economic Missions to Vietnam and Thailand defined malaria control programs as "impact programs." These were programs designed to have a rapid positive effect on local populations in order to build support for local governments and their U.S. supporters.[52] A report of the International Development Administration Board to the president of the United States in 1956 concluded:

> As a humanitarian endeavor, easily understood, malaria control cuts across the narrower appeals of political partisanship. In Indochina, areas rendered inaccessible at night by Viet Minh activity, during the day welcomed DDT-residual spray teams combating malaria. In Java political tensions intensified by overcrowding of

large masses of population are being eased partly by the control
of malaria in virgin areas of Sumatra and other islands, permitting these areas to be opened up for settlement that relieves intense population pressures. In the Philippines, similar programs
make possible colonization of many previously uninhabited areas, and contribute greatly to the conversion of Huk terrorists to
peaceful landowners."[53]

Between 1950 and 1972, various U.S. agencies spent roughly $1.2
billion on malaria control activities, almost all of which employed
DDT or other pesticides.[54]

Local governments also perceived control as an instrument
for gaining popular political support. Thus, as the International
Development Administration Board report noted, "The present
governments of India, Thailand, the Philippines, and Indonesia
among others, have undertaken malaria programs as a major element of their efforts to generate a sense of social progress, and
build their political strength."[55] Similarly, the distinguished Indian malariologist A. K. Viswanathan pointed out, "No service
establishes contact with every individual home at least twice a year
as the DDT service does unless it be the collection of taxes."[56]

Cold War politics shaped the development of postwar malaria
control efforts in another way. It prevented the linking of malaria
control programs with broader efforts at social and economic development, despite the fact that many public health experts, including some of those who supported the predominant role of
pesticides, argued for such a linkage.

The rapid success of malaria control using DDT, combined
with other medical breakthroughs, including the development of
antibiotics, led to a dramatic fall in mortality worldwide. One
of the consequences of this decline was an equally dramatic rise
in population in many developing areas of the world. This led
a number of prominent public health leaders and other observers to question whether disease control programs were actually
making matters worse. One of the most vocal critics of the role
of disease programs in causing overpopulation was neo-Malthu-
sian naturalist William Vogt. Vogt's *Road to Survival,* published in
1948, would influence a generation of ecologists concerned about

overpopulation, including Paul Ehrlich. Vogt questioned whether it was a kindness "to keep people dying from malaria so that they could die more slowly from starvation."[57] He even suggested that malaria was "a blessing in disguise, because a large portion of the malaria belt is not suited for agriculture, and the disease has helped to keep man from destroying it—and wasting his substance upon it."[58]

These attacks led to the development of postwar population control programs. In the meantime, economists argued that disease control programs needed to be linked to broad-based efforts to raise standards of living and to create new economic opportunities—in effect, to develop integrated health and development programs like those of the 1930s.

Gunnar Myrdal was one of the first economists to speak to the question of the economic consequences of health programs. He expressed his views in a speech to the World Health Assembly in 1950:

> The economic value of preventing premature death . . . depends entirely upon whether such an economic development is underway which ensures productive work for the greater number of people we thus keep alive. If the economic situation is stagnant and remains substantially as it was and is in large regions of the world where people live in overcrowded conditions on maltreated land under primitive cultivation the health reforms serve, from an economic point of view, only to speed up a process towards increased relative over-population and aggravated pauperization.[59]

Public health leaders seconded this approach. One of the most prominent postwar leaders in international health was Yale public health professor, C.-E. A. Winslow. In his classic study of the economic benefits of health, *The Cost of Sickness and the Price of Health,* he noted:

> The interrelationships involved make it abundantly clear that the public-health programme cannot be planned in a vacuum, but only as a vital part of a broad programme of social improvement. . . . It is not enough then, for the health administrator to develop

the soundest possible programme for his own field of social endeavor. . . . He must also sit down with experts on agriculture, on industry, on economics, and on education and integrate his specific health programmes as part of a larger programme on social development.[60]

Paul Russell articulated a similar position in his book *Man's Mastery of Malaria*. Russell, a malariologist for the International Health Division of the Rockefeller Foundation, had spent years combating malaria in India, the Philippines, and Latin America. He was a member of the WHO Expert Committee on malaria during the early fifties. The concluding chapter of his book was titled "Malaria Prophylaxis and Population Pressure" and was a response to those who were concerned that disease control programs would lead to overpopulation. In it he argued, "That physicians, malariologists, and sanitarians integrate their activities with those of agriculturalists, demographers, social scientists, economists, educators, political and religious leaders is of the utmost importance. For only thus can there be joint planning of social reorientation that will result not in bigger populations but in healthier communities."[61]

Calls for integrated disease and development programs were short-lived, however. Early efforts to link food programs with disease control, involving a partnership between the Food and Agriculture Organization and the WHO in the early fifties, floundered over disagreements concerning the use of pesticides. Further attempts to integrate disease control with broader development programs fell victim to Cold War politics. This political climate kept social medicine activists from gaining leadership positions in the new postwar international health organizations. Rajchman, whose politics had cost him his position as head of the league's Health Office in the late 1930s, was unable to take on a leadership role in the newly created United Nations Children's Fund (UNICEF), despite having been its prime designer. Similarly, Štampar played a critical role in the development of WHO and yet, as an east European and advocate of social medicine, was blocked from becoming its director. More profoundly, prewar models of integrated rural health and development were viewed with suspicion. The

Italian bonification successes were viewed as Fascist, and the TVA became a symbol of statist approaches to development associated with communist command economies. When President Truman nominated the TVA's director, David Lilienthal, to head up the new Atomic Energy Commission in 1946, his appointment was held up for weeks as senators examined his political record. Senator Taft claimed that he was too soft on issues of communism and the Soviet Union.[62]

The same opposition to large-scale integrated models of development also insured that there was a postwar division of labor within the UN family of organizations. Thus health became the responsibility of WHO; food and nutrition was housed in the Food and Agriculture Organization; economic development in the UN Development Program; education in UNESCO; and labor in the International Labor Organization. This balkanization of development made efforts to promote integrated health and development programs difficult to achieve or, in some circles, even talk about. In the end, the need to integrate malaria control with a wider program of social and economic uplift was viewed by some as politically unacceptable, while others, intoxicated with the power of DDT, saw it as unnecessary.

By the early 1950s house spraying with DDT and other insecticides had become the weapon of choice in the war against malaria. Over the next two decades new drugs, including chloroquine, would be used by individuals to treat and prevent individual cases of malaria. However, public health programs aimed at eliminating malaria would consist of an all-out chemical attack on anopheline mosquitoes. Malaria had become a vector-borne disease.

CHAPTER SIX

Malaria Dreams

In 1957 epidemiologist George Macdonald published *The Epidemiology and Control of Malaria*. Like Angelo Celli's *Malaria: According to the New Researches*, Macdonald's book was intended to provide a model for malaria control. Unlike Celli, Macdonald viewed malaria not as a social problem but as a mathematically definable relationship among mosquitoes, malaria parasites, and human hosts. To explain the dynamics of malaria, he developed a mathematical model, which included a number of variables that affected the transmission rate: the abundance of *Anopheles* relative to human population, the propensity of the vector to bite a human host, the proportion of bites that were infective, the duration of the reproduction cycle that occurred within the mosquito, the probability that a mosquito would survive a single day, and the rate of recovery of the human host. While all of these variables affected transmission, he concluded that the probability of the mosquito surviving a single day was the critical factor, suggesting that malaria transmission could be interrupted simply by shortening the life-span of the mosquito vector, thereby preventing the parasite from developing fully. By doing this, the basic malaria reproduction rate—that is, the chance of a case of malaria producing another case—would be reduced to less than 1. When this happened, transmission would cease.[1] Macdonald's mathematical

150

model provided the theoretical underpinning for a bold attempt
to eliminate malaria as a public health problem.

LAUNCHING THE MALARIA ERADICATION PROGRAMME

In the spring of 1955 the Eighth World Health Assembly convened
in Mexico City at the Palais des Arts. The assembly, which had
met on an annual basis since 1948, when the World Health Organization was founded, was made up of representatives of WHO
member states. The assembly's agenda was set each year by the executive committee and the director general. It normally included
several proposals, which the assembly was asked to vote on and
ratify. These ranged from the establishment of new health programs, to budget requests, to health declarations. In Mexico City,
the assembly was asked to ratify a proposal to eradicate a disease
from the face of the earth. The disease was malaria.

In retrospect, and to many at the time, this proposal seemed
like an act of extreme hubris.[2] By the director general's own estimate 600 million people were exposed to malaria worldwide, not
including the Soviet Union or China. Moreover, earlier attempts
to eradicate hookworm, yellow fever, and yaws, which were much
less prevalent than malaria, had failed. How was it possible that a
disease of such magnitude could be eliminated? The answer lay in
three letters: DDT.

In presenting the case for global eradication, the director general noted the following facts. First, malaria was a major source of
disability as well as a drain on economic development over large areas of the globe. Second, health workers operating in various parts
of the globe had had tremendous success in employing long-acting
(also know as "residual") pesticides, particularly DDT, to control
malaria. Third, evidence from several parts of the globe, including Italy, Venezuela, Crete, Ceylon, and Mauritius, demonstrated
that malaria eradication by residual spraying was both economically and technically feasible. By accelerating existing DDT control methods, it would be possible to eliminate malaria as a public
health problem and contribute to economic development. Vector
control had moved from being a contested method of malaria control to a weapon that could now be used to eradicate the disease.

The eradication strategy was based on the use of DDT to kill female anopheline mosquitoes and interrupt malaria transmission following Macdonald's mathematical models. If transmission could be prevented for several years, then the parasites residing within previously infected individuals would disappear. At that point spraying could be halted, as there would be no parasites left to infect anopheline mosquitoes. Elimination would take longer where *P. vivax* was the dominant parasite because of its ability to persist for long periods within the human host. The more lethal *P. falciparum,* which runs a more acute course, could be eliminated within only one transmission season.

The strategy, developed by expert committees at WHO, involved four stages. The first was a period of *preparation,* during which the epidemiological characteristics of the disease were identified and the attack plan developed. An *attack* phase, involving the use of indoor residual spraying to eliminate transmission, followed. The walls of every house or hut in which humans slept were to be regularly sprayed with DDT. Underlying this tactic were two assumptions: that female anopheline mosquitoes fed primarily indoors; and that these same mosquitoes rested on the walls of dwellings after they took their blood meal from a sleeping inhabitant. The resting behavior brought them in contact with the DDT. As it turned out, neither assumption was completely valid.

The attack phase was to continue until the number of infective human carriers, based on the screening of infants, was zero or close to it. At this point spraying was halted and the *consolidation* phase began. During this phase, active case finding and treatment was undertaken to eliminate the remaining cases of malaria. Countries that experienced no new cases could be certified as having achieved eradication. However, surveillance needed to be sustained to prevent the reintroduction of cases from neighboring countries. This *maintenance* phase would continue until global eradication had been achieved. It was estimated that efficient eradication programs could achieve completion of the first three phases within five to eight years.

Not everyone was convinced that eradication was feasible. The effectiveness of DDT for controlling, let alone eradicating ma-

Malaria Spray Team in Taiwan

Source: Agency for International Development, *Malaria Meeting the Global Challenge*, A.I.D. Science and Technology in Development Series (Boston: Oelgeschlager, Gunn & Hain, 1985), 43.

laria in many areas of the developing world was unknown. As late as February 1955, the executive board of the WHO, reviewing data on the effectiveness of insecticides in controlling malaria, observed that it was not known whether DDT would be absorbed by mud surfaces, reducing its residual killing effects. One member of the board viewed this as a serious concern, because in most malaria-ridden areas mud-walled houses were the commonest form of dwelling.

Others noted that in many parts of the developing world malaria control programs were poorly staffed and disorganized. This had led to the misapplication of DDT. One of the most vocal and respected supporters of the eradication strategy, Dr. Paul Russell, had firsthand knowledge of the disarray that existed in many spray programs and of the administrative difficulties that stood in the way of eradication efforts. He nonetheless pushed for the adoption of the eradication proposal.

At the time that the eradication resolution was passed, few of those who supported eradication believed that it could be achieved in sub-Saharan Africa in the foreseeable future, even though the overwhelming majority of malaria cases and deaths occurred in this region. Logistical problems, limited public health capacity, and the high vectorial capacity of the primary malaria vector, the *A. gambiae* mosquito, led Paul Russell and George Macdonald to conclude that malaria eradication in Africa would have to wait for years. Few eradication supporters seemed to acknowledge, however, that without a strategy for eradication in Africa, the global eradication of malaria was an unrealistic dream.

No one knew just how much eradication would cost or how the funds would be raised. The WHO estimated that it would cost $600 million for the first five years. This estimate was subsequently increased to $1,806 million in 1963. Many doubted that sufficient funds would ever be available. Most of the countries in which malaria was a problem were poor and unable to fund even normal control programs. Supporters of eradication countered that eradication represented a large up-front capital investment that would be quickly returned by the elimination of long-term recurrent expenses associated with maintaining control measures indefinitely. It was also argued that eradicating malaria would remove a major barrier to economic growth in the developing world.

Underlying all of these concerns was the question that public health officials had debated since the beginning of the century: could eradication be achieved in underdeveloped, resource-poor environments? Or, put in the terms of the earlier debate, could malaria be eliminated without simultaneous social and economic development?

Finally, and perhaps most importantly, there was growing evidence that vector resistance to DDT and other pesticides was increasing, threatening the ability of health authorities to keep malaria under control. However, rather than raising concerns about the feasibility of the proposal, this fact served to further justify the launching of an eradication campaign. Eradication had to be executed before the central weapon against malaria was no longer viable. The director general concluded that "There is . . . no other logical choice: malaria eradication is clearly indicated, presents a unique opportunity and should be implemented as rapidly as possible. Time is of the essence."[3] The global strategy for the eradication of malaria was thus a race against time in which the WHO was gambling that malaria could be eliminated before vector resistance precluded this possibility. It turned out to be a poor bet.

Other forces were at play behind the stated justifications for eradication. First, eradication strategies had already been adopted by regional health organizations, including the Pan American Health Organization and the South East Asian Regional Organization. The World Health Assembly risked undermining the WHO's leadership role in international health if it chose to reject the proposal. Second, the early success of insecticides in greatly reducing the toll of malaria worldwide had produced a growing sense of complacency among government officials and funding agencies regarding their need to support malaria programs. As early as 1948 Russell noted that there was a growing concern within the U.S. National Institutes of Health (NIH) about the future status of malaria research. At a meeting in January of that year discussions of malaria research centered on two questions. "To what extent is malaria research in jeopardy because of the present manner of thinking on the part of budgetary authorities?" "What justification is there for continued research in malaria and what facts could be presented in support of such justification?"[4]

In September participants at a Malaria Section Meeting of the NIH discussed the director's proposal to merge six sections into three and join malaria with tropical diseases. Russell noted in his diary, "The tendency is to consider malaria a finished story, but actually now is the time to give this disease a very great emphasis."[5]

Fred Soper, too, recognized the threat that bureaucratic complacency represented for the war against malaria in the Western

Hemisphere. In his statement advocating a coordinated malaria eradication program for the Americas, he noted "the appearance of another type of resistance—no less serious—in officials in charge of public funds who seem less and less inclined to increase or even maintain allotments for a campaign that, both socially and politically, seems to have lost its timeliness." He went on to stress that it was "of the greatest importance to eradicate all sources of infection before this type of resistance has become deeply rooted."[6]

The very success of DDT in controlling malaria threatened to undermine what had been achieved. For the primary advocates of eradication—Soper, Russell, Arnaldo Gabladon, and E. J. Pampana—who had devoted much of their careers to developing programs for controlling malaria, this must have been a troubling prospect at several levels. As humanitarians, they were genuinely concerned about the consequences that funding cuts would have on the health of populations that malaria control programs had protected. As public health administrators, they were no doubt also concerned about the deterioration of the programs they had worked so hard to build.[7]

Ultimately malaria eradication was buoyed by a growing faith in the ability of Western science and technology to transform underdeveloped countries. This faith was part of what has been called "the culture of development." The attitude of "know how and show how" emerged out of World War II and the myriad of technical innovations, from antibiotics to nuclear weapons, that the war had produced. It was clearly reflected in U.S. president Harry Truman's Point Four Program (for applying U.S. technical "know-how" to international development programs) and subsequent U.S. technical assistance programs. Of perhaps equal importance for this growing faith in technology was the expansion of the social and behavioral sciences, and particularly political science, psychology, sociology, and the newly emerging subfield of development economics. The 1950s saw an emerging self-assurance among social scientists in their ability to predict outcomes and control for a whole range of social, economic, and cultural variables. In the United States, motivational research firms advised manufacturers on buyers' preferences and how they could be manipulated to increase the consumption of particular products.[8] The universalizing claims of formalist social science disciplines made it easy to make the cultural leap from selling laundry detergent to housewives in Toledo to getting Africans to accept new fertilizers, hybrid plants, and packaged baby formula—or take antimalarial tablets. The new culture of development was clearly reflected in the 1951 Department of State's document on the Point Four Program, in which malaria control activities using DDT were described.

> The most dramatic results from the employment of a very small number of skilled men and very small quantities of scientifically designed materials have been achieved in the field of medicine. In many areas of the world one trained public-health doctor or a group of two or three working with local people able to follow their guidance have been able to rout one of man's oldest and deadliest enemies.[9]

The passage not only stressed the ability of Western technology to bring about change but also privileged the skills and knowledge of the outside expert, while placing local populations in a position of dependence in need of guidance and assistance. This construction of the recipients of development was as essential to the culture of development as the belief in technology. Faith in both elements underpinned belief in the feasibility of malaria eradication.

In the end, the World Health Assembly passed the resolution, and the Malaria Eradication Programme was launched in 1955. Over the next 14 years, nearly $1.4 billion dollars was spent on the eradication program. The WHO provided technical advice to countries initiating eradication programs. UNICEF provided countries with material support. In addition, individual donor nations contributed to the special fund set up by WHO for eradication and also provided bilateral aid. The largest contributor was the United States, which donated $490 million. An additional portion of the cost of eradication programs, amounting to $650 million, was borne by local governments.

The governments of malarious countries were not universally enthusiastic about eradication. Some governments questioned the wisdom of investing heavily in malaria when they faced so many other social and economic problems, as well as other health issues.

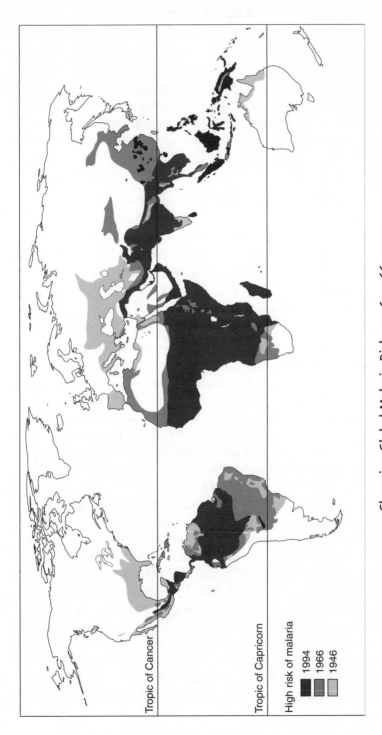

Changing Global Malaria Risk, 1946, 1966, 1994

Source: Jeffrey Sachs and Pia Malaney, "The Economic and Social Burden of Malaria," *Nature* 415, 6872 (2002): 681.

Tropic of Cancer

Tropic of Capricorn

High risk of malaria

1994
1966
1946

This was especially true in countries where existing malaria control programs had significantly reduced transmission. Still other governments questioned the feasibility of the eradication strategy itself. Yet from 1958, access to U.S. support for malaria control depended on a country's adopting the eradication strategy. Those countries not cooperating could face the loss of much of their external support for malaria control. There was thus immense financial pressure to make the transition to eradication. Unfortunately, many countries were not prepared to launch an eradication program, and the precise strategies laid out by WHO were applied imperfectly at best.

ASSESSING THE MALARIA ERADICATION PROGRAMME

The results of the malaria eradication campaign were mixed. By 1970 eradication had been achieved in 18, or 39 percent, of the countries in which it was initiated. An additional 8 countries achieved eradication after 1970, a total of 56 percent. In many other parts of the globe malaria mortality and morbidity were dramatically reduced. However, these declines proved in many cases to be unsustainable, and, as we will see, malaria returned. The areas in which malaria was eliminated correspond roughly to the countries that were affected by malaria in 1946, but not in 1966.

Those nations where malaria was eradicated fell into three categories (see table 1). The first group was made up of economically developed countries located primarily in western Europe and North America. While there was no absolute relationship between economic development and success in malaria eradication, levels of economic development were important. Populations of economically better-off countries, such as the United States, Australia, Singapore, Brunei, and Spain were better able to protect themselves from infection through the construction of screened houses, the use of bed nets, and other forms of personal efforts. Economically stronger countries had more resources to devote to eradication programs, were less dependent on external aid, had a more educated work force to apply eradication strategies, and were more likely to apply the eradication strategies in an effective and timely manner.

The second group of countries to achieve eradication included

Table 1. Countries Achieving Eradication by 1970

Economically Developed	Island Nations	Socialist Countries
Italy	Grenada	Bulgaria
Netherlands	Saint Lucia	Poland
United States	Trinidad Tobago	Romania
Australia	Dominica	Yugoslavia
Brunei	Puerto Rico	Hungary
Singapore	Virgin Islands	Cuba
Portugal	Cuba	
Spain	Jamaica	
	Mauritius	
	Réunion	
	Taiwan	

Data Source: C. Gramiccia and P. F. Beales, "The Recent History of Malaria Control and Eradication," in W. H. Wernsdorfer and I. McGregor, eds., Malaria: Principles and Practice of Malariology (Edinburgh: Churchill Livingstone; New York, 1988, 1360.

island nations located primarily in the Caribbean. Their small size, relative isolation, and ability to restrict the reintroduction of malaria from outside all contributed to their successes.

Finally, the third group was East European countries that were relatively poor, but possessed reasonably well-developed health infrastructures and communications systems. The existence of a health infrastructure was particularly important in the latter stages of eradication, once spraying operations had ceased, because it was essential that any remaining cases be identified and treated. A network of local health centers was needed, and where these did not exist, cases could go undetected, providing a focus for the local transmission of malaria. Where transmission reoccurred, spray operations had to be reinstituted. This prolonged the attack phase of eradication and in turn increased program costs as well as the risk that vector resistance to pesticides would develop. When the WHO declared an end to the Malaria Eradication Programme in 1969, a third of the countries in Latin America, Asia, and the eastern Mediterranean were still in the attack phase. Some of these countries had been spraying for more than 10 years. The prolongation of eradication contributed to disillusionment and bureaucratic resistance to continued support for the eradication strategy within individual countries.

Good communications networks were also important for insuring that spray teams and supplies reached spraying sites in a timely manner. Eradication efforts faltered where spraying operations did not occur early enough in the anopheline breeding season, or where they were incomplete, leaving remote areas of a country uncovered.

Intranational comparisons also reveal the importance of overall social development for the success of eradication. Tropical and nontropical areas of Mexico, such as Veracruz, that had implemented sanitary engineering projects and improved public services prior to eradication were more easily freed of malaria transmission than poorly developed areas with subsistence economies, regardless of the level and frequency of insecticide saturation and drug treatment.[10] Similarly, the Indian state of Kerala was able to eradicate malaria in a relatively short time, while the disease continued to plague other areas of India. This difference reflected Kerala's prior development of an efficient health system together with its heavy investments in education and other social services.[11] Unfortunately, the failure of eradication in India as a whole inevitably meant that eradication could not be sustained in Kerala, because infected persons were free to enter the state.

Climate and epidemiological conditions also appear to have been important in determining the success or failure of eradication. Most of the areas in which eradication was achieved were located in temperate or subtropical regions of the globe where malaria was unstable, meaning transmission occurred intermittently. (Tropical regions that were successful were either islands or economically well-off nations, for example, Singapore and Brunei.) Areas of unstable malaria had two advantages in terms of control. First, there was a close correspondence between infection and disease, because populations did not develop immunities to the disease. It was therefore easy to identify infected cases. By contrast, in areas of stable malaria, immunity levels were high, and there were substantial asymptomatic carrier populations. The existence of large numbers of asymptomatic carriers presented a serious challenge to surveillance efforts in the latter stages of eradication, when it was essential to identify every infective case. Second, a natural reduction of vectors occurred in temperate areas during colder periods, which made reducing vector populations easier. Finally, the types of vectors present in temperate regions of the globe also tended to be less efficient in transmitting malaria than those in tropical areas, which meant that malaria reproduction rates—the number of cases that would be produced by a single case—were generally lower.

Still, it is difficult to disaggregate the effects of climate and epidemiology from those produced by social and economic conditions. While eradication was seldom achieved in so-called tropical countries, these same countries were economically disadvantaged and had poorly developed health infrastructures and limited resources to carry out eradication.

A great deal has been written about why malaria eradication failed to achieve its goals in many countries, as well as why malaria returned in countries that had come close to achieving eradication. Much of the literature has focused on technical and financial problems: the organization of spraying programs, the lack of sufficient supplies of DDT, the development of pesticide and drug resistance. Yet we must not lose sight of the larger social, economic, and political forces that contributed to these problems.

RESISTANCE TO PESTICIDES AND ANTIMALARIAL DRUGS

Malariologists have pointed frequently to the twin problems of pesticide and drug resistance as major obstacles to the successful completion of the eradication program in many parts of the globe. However, it is important to examine the causes of this resistance.

Pesticide Resistance

Pesticide resistance had already begun to emerge before the eradication program was launched in 1955. It was, as we have seen, one of the justifications for attempting eradication. Yet the numbers of malaria vectors that developed resistance to pesticides once eradication had begun increased dramatically. Pesticide resistance was the result of the selective pressure that was applied to mosquito populations through repeated exposure to pesticides. Over time, genetic variation in the exposed population gave rise to resistant mosquitoes. As spraying continued, the resistant strain was able to survive and reproduce, while those that were susceptible were killed off. Eventually, the percentage of resistant mosquitoes exceeded those that remained susceptible. In this sense resistance was inevitable.[12] The development of pesticide-resistant anopheline mosquitoes was accelerated, however, by patterns of pesticide use. Within poorly funded control programs, where spraying operations were incomplete, or where pesticides were watered down to allow for increased coverage, only the most susceptible mosquitoes were killed off, leaving more resistant species to survive and reproduce.

The concurrent use of DDT for malaria control and crop protection also contributed to the development of pesticide resistance in South Asia and Latin America, by greatly increasing selective pressure on local anopheline mosquitoes. In El Salvador, DDT was employed to clear malaria from the Pacific coastal plains. This made possible the rapid expansion of cotton plantations along the coast. Plantation owners eager to increase cotton exports employed DDT to protect their crops from agricultural pests. By 1958 entomologists reported that the local malaria vector *Anopheles albimanus* had lost its susceptibility to all major organochlorine compounds, including DDT. Four years later researchers in southern Mexico reported the same phenomenon in association with cotton growing. In India, widespread resistance was discovered among the two important malaria vectors, *Anopheles culicifacies* and *Anopheles fluviatilis,* particularly in areas in which high-yielding forms of agricultural production using pesticides had recently been introduced.[13]

By 1969, the year in which WHO abandoned the eradication strategy, 56 species of anopheline mosquitoes were reported to exhibit resistance to DDT. Widespread anopheline resistance led public health officials to turn to other organophosphate, carbamate, or pyrethroid insecticides. While some of these proved to be effective in controlling malaria, they were more costly than first-line products. The cost of malathion, which was adopted widely in response to DDT resistance, was 5 times the cost of DDT. Proxupor, another DDT substitute, cost 20 times more than DDT. The increased cost of vector control forced some programs to adopt cost-cutting strategies—reducing areas sprayed or the frequency of spraying. These measures limited the effectiveness of spray operations and, in a number of countries, contributed to increases in malaria transmission. A recent study of malaria control programs in 21 Latin American countries showed that as house spraying declined below effective levels, the annual parasite index rose dramatically.[14]

It is often claimed that environmental lobbyists who pushed for the banning of DDT following the publication of Rachel Carson's *Silent Spring* in 1962 undermined malaria eradication. Environmental concerns did lead developed countries to ban the use of DDT within their borders and limited its production in the 1970s and 1980s. These restrictions, in turn, placed a serious financial burden on control programs in poor countries in which the pesticide was still the most effective and cheapest form of malaria control. Yet the impact of environmental protests and restrictions on the use of DDT came after the Malaria Eradication Programme had largely wound down.

Antimalarial Drug Resistance

Malariologists have also implicated antimalarial drug resistance in the failure of eradication efforts in tropical and subtropical regions of the globe. Drug-resistant parasites exhibit decreased sensitivity to specific antimalarial drugs. This decrease in sensitivity ranges from partial (RI) to complete (RIII). Patients with diminished sensitivity to antimalarial drugs fail to clear their parasites and suffer recurrent symptoms. They also remain a source of infection to malaria vectors. Resistance to chloroquine—a cheap and effective first-line drug for the prevention and treatment of malaria worldwide—was a major blow to control programs. This was particularly true for chloroquine-resistant forms of *P. falciparum,* the most lethal malaria species. As chloroquine resistance grew, so did malaria mortality. In addition, the need to move to second-line and often more costly drugs increased the costs of eradication. Again it is important to examine why resistance developed.

Chloroquine-resistant forms of *P. falciparum* resulted from the selective pressure that was applied to malaria parasites through repeated exposure to chloroquine, in the same way that pesticide-resistant mosquitoes evolved in response to repeated exposure to DDT. These resistant forms of *P. falciparum* had a comparative survival advantage in the face of the subsequent exposure to chloroquine. This advantage was not simply one of survival. The rate of maturation of merozoites, which invade red blood cells, appears to be faster in chloroquine-resistant strains of *falciparum* than in nonresistant strains. Thus, in a person infected with both resistant and nonresistant forms of *P. falciparum,* the resistant forms are likely to propagate faster, invading more red blood cells. Chloroquine-resistant strains of *falciparum* have also been shown to complete the sexual reproduction stage faster than nonresistant forms within certain species of malaria vectors.[15]

Strains of *P. falciparum* resistant to chloroquine were first suspected in Thailand in 1957 and found in patients in Colombia and Thailand in 1960. Since then, resistance to this drug has spread widely and there is now high-level resistance to chloroquine in South Asia, Southeast Asia, Oceania, the Amazon Basin, much of subtropical Africa, and some coastal areas of South America.[16] Resistance to other antimalarial drugs has also emerged over the past 10 years.

As in pesticide resistance, the ways in which antimalarials have been administered have accelerated the emergence of drug-resistant parasites. Efforts to protect populations from malaria through the application of mass drug administration programs in the 1950s and 1960s may have produced the earliest cases of chloroquine resistance. These programs distributed chloroquine as a prophylactic drug, either directly or indirectly in chloroquinized salt, to thousands of people in Southeast Asia, East Africa, and Latin America. As a consequence, large populations, many of whom were already infected with malaria, received dosages of chloroquine that were not sufficient to eliminate the parasites from their blood. This provided the opportunity for resistant strains to survive. There is a strong correlation between the geographical points where mass drug administration programs were initiated and the places where chloroquine resistance first emerged globally.[17] Also contributing to developing resistance were the widespread availability of chloroquine in shops and private pharmacies and the lax regulation of how the drug was used. As I witnessed in Mulanda, consumers with limited resources often chose to reduce the dosage of chloroquine to save money. Some shops even sold subtherapeutic dosages of the medicine.

A closer look at the genesis of chloroquine resistance along the Thailand-Cambodia border in the 1950s and 1960s reveals a great deal about the social-biological dynamics of antimalarial drug resistance. Palin, a small town located in a densely forested area of Cambodia, is a center of gem mining, producing blue sapphires and rubies. It was also ground zero for the development of malaria drug resistance in Southeast Asia.

The mining industry in Palin attracted a continual flow of newcomers from neighboring regions of Cambodia, Thailand, Vietnam, and Burma (Myanmar) as well as from Bangladesh. Between 1,000 and 1,200 migrants arrived each month. As in other parts of Asia, Africa, and Latin America, mining was a way for impoverished farmers to supplement the meager incomes they earned from agriculture. In Palin miners received three to five dollars from local merchants for each gem they unearthed.

Many of the migrant workers came from areas in which malaria was much less intense, and it was estimated that 80 percent of these workers had no resistance to the disease. The mining activities created hundreds of shafts, which collected water from seepage and rains. These created breeding sites for a highly efficient malaria vector, *A. dirus,* which bred in very high numbers. The miners, who stayed for three to four months, lived in the open air and slept under very rudimentary shelters. They were thus exposed to multiple infective bites. The convergence of a highly efficient vector, a nonimmune population, and mining conditions that encouraged both vector breeding and malaria transmission fueled recurrent epidemics of malaria, which apparently started in the 1940s or early 1950s; 80 percent of the cases were caused by *P. falciparum.*[18]

Public health authorities depended on chloroquine to control malaria in Palin because *A. dirus* did not rest inside human habitations. This made domestic spraying with DDT ineffective. The value of chloroquine was compromised, however, by the immunological status of the mine workers, the method by which the drug was distributed, and the continued influx of new groups of workers.

The nonimmune migrants who were exposed to malaria suffered heavy infections and severe clinical symptoms. They consequently required high doses of chloroquine to eliminate the parasites in their blood. Unfortunately, public health authorities in Palin employed a mass drug administration program through which chloroquine was handed out to mine workers in subcurative doses as a prophylactic. This was done twice a year from 1955 to 1962, and weekly during 1958 and 1959. Between 1960 and 1962, antimalarials were administered indirectly through medicated salt. Predictably, the repeated application of subcurative concentrations of chloroquine among a highly infected population eliminated only the most sensitive parasites. More resistant strains survived. This set the stage for the emergence of chloroquine resistance to *P. falciparum.* Resistance was further promoted by patterns of self-medication on the part of miners, who often could not afford the dosages of antimalarials needed to sterilize the blood. Finally, the transmission of resistant strains of *falciparum* to new waves of nonimmune workers, month after month, and their treatment with high but often noncurative doses of antimalarials amplified resistance to the point where the drug became nearly ineffective against *falciparum* malaria. By 1973, 90 percent of falciparum cases were resistant to chloroquine; 70 percent exhibited RIII resistance. In the end, the development of chloroquine resistance in Palin was generated by patterns of economic development in the region, combined with conditions of employment in the mining areas, and the ways in which public health officials administered the drug. Chloroquine resistance was not the inevitable result of natural selection.

From the Thai-Cambodian border, falciparum resistance spread to surrounding areas along with the returning migrant workers. Secondary patterns of dispersal from these surrounding areas contributed to the wider dissemination of chloroquine resistance throughout South and Southeast Asia.

TACTICAL PROBLEMS

Although DDT and drug resistance contributed to the failure of the global Malaria Eradication Programme, there were other more fundamental problems. The program was handicapped from the beginning by its reliance on a single strategy administered from the top down, with little input from below or effort to adapt the eradication strategy to local conditions. In this sense, it violated all that had been learned about vector control over the previous 50 years. As early as 1911, Ronald Ross had demonstrated that the application of multiple strategies was more effective in limiting transmission than the use of a single strategy.[19] This conclusion had become gospel among malariologists over the following 40 years. It was also recognized that all malaria problems were local and that control strategies had to be based on a thorough understanding of local ecologies.

Yet the Malaria Eradication Programme exhibited little flexibility in adjusting its methods to local circumstances or in trying to understand how local conditions might affect eradication outcomes. In particular, little effort was made to understand local reactions to spraying, or how local social conditions and cultural attitudes might affect how local populations reacted to spray operations. This hampered the successful completion of spray operations in many rural areas. For example, in some areas women left their homes early in the morning and did not return until evening. While gone they locked their huts. This prevented sprayers from gaining access. In other areas, people found the odor produced by the mixture of DDT and kerosene objectionable and replastered their walls after the spray teams had passed through. The simple act of spraying the walls of huts and houses could be very disruptive: people had to remove their belongings from their homes, including anything that was hanging on the walls. Not everyone was cooperative. Where resistance occurred, spray teams sometimes employed force to make sure that all dwellings were sprayed. Resistance often reflected the failure of the eradication program to involve local populations in the planning and execution of spray operations. The situation was not helped by the fact that in many areas spray teams arrived with little explanation or warning.

More problematic still was the failure of the eradication program to develop alternative methods to deal with populations, like those in the mountains of Cambodia, that lived in shelters that provided no surfaces for spraying. Nomad populations created yet another set of problems for a program wedded to spraying the walls of human habitations with DDT. There were also no ready solutions for the problem of malaria where the primary vectors did not feed or rest indoors. All of these local problems created gaps in the coverage of spray programs, making the eradication of malaria impossible to achieve.

In addition, the eradication program provided few resources to support continued research on the problems of malaria and its control. In fact, faith in the ability of DDT to eliminate malaria made malariology appear to be superfluous. It has been said that while malaria eradication failed to eradicate malaria, it succeeded in eradicating malariologists.[20] When the single strategy upon which eradication was based began to falter, there was little in the research pipeline that would either help explain the problems eradication faced or provide the basis for an alternative set of strategies.

FINANCIAL CONSTRAINTS

The Malaria Eradication Programme also failed to achieve its goals because it lacked the resources needed to carry the campaign to its conclusion in resource-poor tropical countries. Many of the logistical problems that countries faced—lack of equipment and transport, scarce supplies of pesticides, insufficient numbers of trained personnel—reflected inadequate funding. Both bilateral and multilateral support for the program waned during the early 1960s. UNICEF, which had contributed between $4.1 and $8.8 million annually, began phasing out support in 1964. It also reduced the number of staff members working on malaria by half between 1961 and 1969. Voluntary contributions to the WHO's Malaria Eradication Special Account peaked at $4.8 million in 1964 and then dropped off to less than $1 million. The contribu-

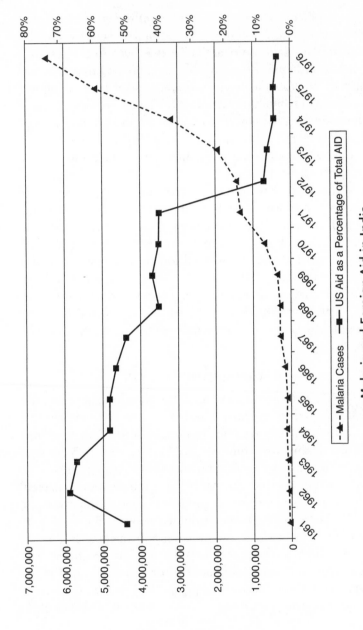

Malaria and Foreign Aid in India

- - ▲ - - Malaria Cases —■— US Aid as a Percentage of Total AID

Sources of data: Government of India, Central Bureau of Health Intelligence, "Positive Cases and Deaths of Malaria in India 1961 to 2004," in Public Health Statistics, *Health Information of India* (2004), chap. 10, www.cbhidghs.nic.in/hii2004/content1.asp.

tions of the United States, by far the largest bilateral donor, also peaked in the early sixties.[21]

Overly optimistic assessments of how long eradication would take and how much it would cost led to donor fatigue. Some have argued that had eradication been achieved in the time frame that was projected there might have been just enough money. But as eradication dragged on in one country after another, the available funds proved to be insufficient and donors were hesitant to throw more money at the problem. In countries that succeeded in eradicating malaria, the average cost of achieving protection was estimated at US$2 per person—but the original estimates had put this figure at 11¢ per person. The costs of maintaining surveillance once eradication had been achieved were also high.[22]

The decline in funding was not entirely based on disenchantment with malaria control. The United States withdrew funds from specific countries for political reasons. For example, it cut foreign aid to India in 1972 in response to India entering into a friendship treaty with the Soviet Union during its armed conflict with neighboring Pakistan. Similarly, socialist tendencies by the Sri Lankan government led the United States to withdraw foreign aid from that country in 1963. At the time Sri Lanka had all but eradicated malaria, with 6 reported cases in the entire year. Over the next five years, malaria exploded upward, reaching 1 million cases by 1968.[23]

Countries involved in protracted efforts to eradicate the disease were increasingly dependent on their own resources. Where financial resources were limited, or when so many other problems needed attention, resource-strapped countries began to question the value of large investments in malaria control, financial support for malaria declined, and campaigns failed. It is also worth noting that in many of these countries, malaria control programs had succeeded in eliminating malaria as a problem for the economically well-off and politically powerful. Malaria, in effect, had settled in among the poor and powerless. The marginalization of malaria together with the rising cost "per-case-prevented" as the completion of the program approached, discouraged local governments from sustaining high levels of support for eradication measures.

LACK OF ECONOMIC PAYOFF

Disenchantment with lack of progress across a broad range of developing countries raised questions about the feasibility of the eradication strategy. Added to this disenchantment was the failure of the Malaria Eradication Programme to demonstrate that eradication would contribute to the economic development of formerly malarious countries. This had been a central justification for eradication in 1955. The director general of WHO, M. G. Candau, attempted to demonstrate the economic value of eradication, employing an Indian economist to produce a study on the subject. The project ended in failure, as no one could provide data to support the assertion. Writing to Yale economist A. M. Payne in 1962, Candau noted, "On several occasions we have attempted to produce solid proof of the relationship and express it either in £.s.d. or Dollars. Again and again we have noticed that this is a much more difficult subject than we had anticipated and all our preliminary documents were considered unfit for publication."[24] To make matters worse, a number of economists and demographers took up the question of the economic benefits of disease eradication in the 1960s. They raised a series of methodological and substantive questions that challenged assumptions about the economic benefits of eradication.[25]

In a number of regions, eradication actually had a negative impact on the economic well-being of local populations. For example, the eradication of malaria in the low-lying areas of Nepal opened the region to settlement by upper-class Hindus from the foothills of the Himalayas, who had formerly avoided the region because of malaria. The consequences that this invasion had on the original inhabitants of the lowlands were described by an American physician working in Nepal in the early 1990s:

> When the Tharu were exclusive to the region, they cultivated crops in excess of their consumption. The villagers sold the surplus to the inhabitants of the hill and mountain regions. This changed after the eradication of malaria; the upper castes swarmed upon the plains taking not only business away from the Tharu but also monopolizing most of the Tharus' land. Today many Tharu are working as bonded laborers on land they previ-

ously owned. The men and women who formerly sold crops are now unable to obtain enough food to feed their families. Dietary deficiencies, political marginalization, and lack of resources also exacerbated mental illness.[26]

Similar patterns of displacement, following the elimination of malaria occurred along the Pacific coastal plain of El Salvador and in the Pongola Valley of South Africa. These cases showed that the assertion "malaria control leads to economic development" needed to be followed by the question "Development for whom?"

PULLING THE PLUG ON ERADICATION

With support waning and costs rising, the World Health Assembly revisited the program and concluded in 1969 that while malaria eradication remained its ultimate goal, programs should shift back to control strategies. What this meant in practical terms was unclear. For the first few years following the termination of the eradication campaign there was little guidance from WHO about how countries should conduct control programs. In many, if not most countries the shift to control simply meant business as usual, continuing the eradication strategies, only with less financial support. This often led to wasted efforts and resources. Programs continued the costly strategy of trying to track down every case, even though control, unlike eradication, did not require this degree of perfection.

In the 1970s WHO advisory panels on malaria recommended that malaria control programs be integrated into primary health care activities. This was both a cost-cutting move and a victory for those within WHO who had argued against the development of large-scale vertical programs in favor of investments in the development of primary health care systems. What the policy meant in practice varied from country to country. Many developing countries had no well-developed system of primary care, and in these countries there was little in the way of a health infrastructure into which malaria control activities could be integrated. Malaria workers found themselves serving many functions unrelated to malaria control and were prevented from fulfilling their control responsibilities. Even where health services existed, integrating malaria control programs into these systems caused the centralized organization necessary for coordinating spraying operations to be downsized or dismantled. In most places, integration weakened malaria control programs.

WHO's decision to halt the eradication program led donor countries to cut back further on their support of malaria control. While foreign assistance for antimalarial programs between 1957 and 1967 had totaled $1.4 billion, total assistance in the eight years following the end of eradication totaled only $250 million. WHO's malaria advisory staff decreased progressively from 444 to 155, and UNICEF's staff dropped from 115 to 37 between 1967 and 1977. Multilateral and bilateral support in commodities and services dropped from $16.4 million to $6.3 million annually during the same period. Considering the devaluation of the dollar, this represented a drop of 80 percent.[27]

At the same time, new health problems were pushing their way on to the agendas of both multilateral and bilateral aid organizations. Smallpox eradication, initially proposed in 1958, was finally initiated in 1966. There was also growing concern about population growth and increasing international aid investments in family planning programs. Both sets of initiatives drew resources away from malaria. Foreign assistance for health programs provided by the U.S. Agency for International Development shifted dramatically from malaria to family planning in the late 1970s, and local governments followed suit. They could ill afford to waste resources on a problem that appeared to be no longer a priority of international donors, especially where so many other problems existed.

Declining support for malaria control forced many country programs to cut back on, and in some cases close down, control operations. As control programs deteriorated, the improvements in malaria morbidity and mortality that had been achieved under the Malaria Eradication Programme were reversed—sometimes in dramatic fashion. In the highlands of Madagascar, where spraying operations with DDT had eliminated malaria by 1960, the government continued irregular spraying operations to prevent reoccurrences up through 1975, and malaria centers identified and treated cases that occurred. These centers were closed, however, in

1979, and by 1980 the government had ceased all control operations, believing that the disease was no longer a problem in the
highlands. Over the next six years, cases began to occur in various
parts of the highlands creating new pockets of malaria transmission. Without malaria control programs in place, the new foci
went unchecked. Beginning in 1986 these pockets provided the
starting points for a series of major epidemics that tore across the
highlands. Having been protected for decades, the highland population had no resistance to the reemerging disease. During the
worst of the epidemics, tens of thousands of people died in one
three-month period.[28] Malaria had returned to the Madagascar
highlands with a vengeance.

Nearly everywhere malaria eradication had not been achieved,
malaria rebounded during the 1970s and 1980s. In India, where
malaria had dropped from an estimated 75 million cases in 1947 to
only a few thousand in 1960, malaria cases increased dramatically
from the early 1970s, reaching 2 million by 1989. While this number did not come close to preeradication numbers, malaria had
once again become a serious public health problem over wide areas
of the country. In Southeast Asia there were an estimated 110 to 115
million total cases among the Southeast Asia Regional Organization countries prior to eradication. This number had been reduced
to around 100,000 cases by 1965. By 1975 the total had risen to
just over 7 million cases. The number of annual cases remained at
roughly this level through the end of the century.[29] In sub-Saharan
Africa, where malaria had been brought under control only in
southern Africa, it is estimated that the number of cases from 1982
to 1995 was four times as high as the number from 1962 to 1981.[30]
According to the WHO, worldwide malaria mortality reached its
lowest point in the early 1960s and has grown steadily ever since.[31]
Much of this rise in mortality reflected a sharp increase in mortality in sub-Saharan Africa. But increases have also occurred in
several other areas.

In the wake of the failure of malaria eradication, public health authorities concluded that reliance on DDT and killing mosquitoes
to eliminate malaria had been a mistake. This was an important
lesson. But they did not learn a larger lesson. Economic and political

forces that DDT could not eliminate undermined eradication. Economic underdevelopment and the lack of well-developed health and communication systems in many tropical regions thwarted eradication efforts. The poverty of many of the countries in which malaria eradication was started prevented those countries from sustaining control once international support waned. The ill-advised use of pesticides for both public health and economic purposes encouraged pesticide resistance. Rural poverty drove farmers in Southeast Asia to seek income mining for gemstones on the Thai-Cambodian border, where they were exposed to malaria and contributed to the emergence and spread of chloroquine resistance. The Cold War politics that had undermined earlier efforts to develop integrated health and development programs following World War II, led the United States to withdraw financial aid from India and Sri Lanka at critical moments in their control efforts, contributing in both cases to a resurgence in malaria. More importantly, as we will see in the next chapter, when the spraying ended and control measures were relaxed, societal forces once again contributed to a rising tide of malaria.

CHAPTER SEVEN

Malaria Realities

In August of 1989, a 52-year-old banker living in a wealthy, red-tile-roofed subdivision in the town of Rancho Santa Fe in northern San Diego County developed a high fever and chills. The man was subsequently treated for *P. vivax* malaria. He had not traveled outside the United States. Four days later, a migrant worker from Mexico, who worked as a gardener in the banker's community, collapsed on the grounds and was taken to the hospital. He too was treated for malaria. The migrant lived in a camp made up of makeshift shelters constructed from plastic tarpaulins and cardboard materials with some 40 other migrant workers. Three of the workers subsequently tested positive for malaria. The camp was located one mile from the banker's subdivision.

The banker and the migrant workers were linked by the transmission of malaria, most likely conveyed by an *Anopheles hermsi* mosquito, which bred in a stream next to the migrant's camp, below the subdivision in which the banker lived. According to the Centers for Disease Control, the mosquito was in all likelihood infected by a migrant who had arrived in the United States carrying a vivax infection. Yet, in a broader sense, the banker and the migrants were connected by the history of differential global economic development that had produced very different standards of living and health on the two sides of the U.S.-Mexican border. While poverty, as well as concentrations of wealth, existed on both sides of the border, the disparity between poverty and wealth was, and is, much greater south of the border. As a result, tens of thousands of migrants, many of them illegal, made their way annually across the border in search of jobs that paid wages higher than were available in Mexico.

Many others, who traveled from as far away as Central America, joined these Mexican migrants. Once inside the United States, they easily found employment in the agricultural, construction, and landscaping industries. Most stayed from May to October, some longer. Working for low wages, without benefits or the provision of housing, and in many cases fearful of being forced to move by local authorities or picked up by Immigration and Naturalization Service agents, they constructed camps hidden away in narrow valleys, near their places of employment. When public health workers tried to trace cases related to the migrant who had collapsed from malaria in Rancho Santa Fe, they could not see the camp until they were within 10 feet of it.

These Hispanic migrants resembled the hill town residents who traveled to work on agricultural estates of southern Italy in the nineteenth century and the poor farmers of the Sertão who sought work on coastal plantations of northeastern Brazil in the 1930s. While many of them tended lawns and shrubs rather than fields of grain or sugarcane, the Hispanic migrants were nonetheless part of a system of agricultural production that depended on the migration of impoverished rural workers. They were suburban agriculturalists in a globalized economy that placed them at risk of infection.

The 1989 epidemic was one of a series of malaria epidemics involving the transmission of vivax malaria that occurred within San Diego County between 1986 and 1990. In each of these outbreaks, migrant workers provided the initial index cases from which transmission began. Like the Archangel epidemic of 1922, the malaria outbreaks in San Diego County were also part of a larger regional malaria epidemic. They were, in fact, the northern extension of a growing epidemic of malaria in Mexico and Central America. This broader resurgence of malaria resulted from the collapse of efforts to control malaria by residual house spraying with DDT, following the unsuccessful effort to eradicate the disease between

1956 and 1982. In Mexico, the number of new cases rose from just over 20,000 to nearly 150,000 between 1982 and 1986. Thus, the banker and the migrants were both victims of the global resurgence of malaria.[1]

MALARIA AND AGRICULTURAL DEVELOPMENT

The global expansion of malaria since the mid-1960s can be explained in part by declining international support for efforts to combat a disease that seemed no longer a direct threat to the health of developed northern nations. Even in subtropical and tropical countries, malaria had become a less important problem as it settled in among the poor. Complacency resulting from early successes, or disillusionment caused by the failure of eradication, also led governments to curtail malaria control programs.

But it would be simplistic for us to ascribe the resurgence of malaria to a lack of money, or to complacency, or disillusionment alone. If the earlier history of malaria's expansion and withdrawal tells us anything, it is that the epidemiology of malaria has always been shaped by broader societal forces. These forces have created breeding opportunities for malaria-carrying mosquitoes, exposed human populations to infection, and contributed to the movement of malaria parasites. Not surprisingly, the changing global incidence of the disease since the 1970s has been closely linked to patterns of global political, economic, and social change.

As in the past, changing patterns of agricultural production, including the development of the migrant-worker-based farming system of the southern United States, continue to play an important role in the fortunes of this "tropical disease." The link between agricultural development and malaria can be seen in the recent histories of Swaziland, Brazil, and El Salvador.

Agricultural Production in Swaziland

Situated on the border between South Africa and Mozambique, Swaziland experienced a pattern of malaria similar to that described in the eastern Transvaal in chapter 4. Malaria was hyperendemic throughout the lowveld but declined in frequency as one moved westward into the higher regions of the country. This pattern was occasionally disrupted by major regional epidemics that

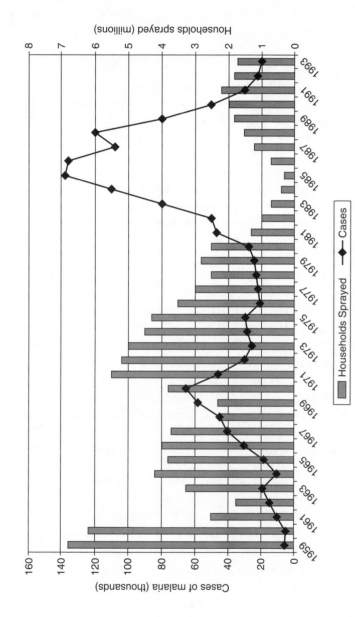

Malaria Cases and Insecticide Sprayings in Mexico

Source: Centers for Disease Control and Prevention, Malaria Control Programs, Mexico, www.cdc.gov/malaria/control_prevention/mexico.htm.

spread over much of the country.[2] As in South Africa, these regional epidemics were directly linked to patterns of colonial development. The presence of malaria prevented European settlement and the development of agricultural estates in the lowveld before World War II.

In the 1940s malaria control operations using DDT opened up the lowveld for extensive cultivation as it had in India and Central America.[3] The Colonial Development Corporation targeted the region for the expansion of sugar and citrus production. To make this possible, malaria control units were established and indoor residual spraying began on a limited basis in 1949. Spraying was expanded to cover all of the lowveld during the 1950s. After only three years of spraying, parasite rates in children living in sprayed areas dropped from 65 to 2 percent. From 1956 spraying was slowly withdrawn from those areas in which no new cases had been reported in the previous two years, and by 1958 only a 15-mile barrier between Swaziland and Mozambique, where no control measures had been instituted, was being sprayed. Entomological work in 1958 and 1959 concluded that *A. gambiae,* the primary vector in Swaziland, had almost totally shifted its living feeding patterns, resting out of doors and feeding off animals. But it was later recognized that in Swaziland *A. gambiae* was in fact a complex of three subspecies of anopheline mosquitoes and that what had happened was not a shift in the mosquitoes' behavior, but the elimination of the subspecies that fed indoors on human hosts. The number of new cases in Swaziland dropped from nearly 8,000 in 1946 to around 100 per year in the late fifties and early sixties. While recognizing the potential threat of malaria's reintroduction in the lowveld by migrants from Mozambique, and the need to maintain surveillance, medical authorities in Swaziland stated that malaria had been just about eradicated by 1959.

By the late 1970s, however, the annual number of cases had risen again to more than 1,000 per year. This turnaround, as in Central America and India, was related to the expansion of agricultural production in the lowveld. Unlike the situation in India and Central America, however, the resurgence of malaria in Swaziland was not caused by the development of vector resistance.

Instead, conditions of production created opportunities for the reintroduction and expansion of the disease.

The successful control of malaria in the lowveld permitted the Colonial Development Corporation to develop irrigation projects for the production of sugarcane. Cane production got underway in the late fifties, and malaria transmission began to expand soon afterward. An initial flare-up of indigenous cases occurred in 1960 in and around the sugar estates, and larger outbreaks occurred in 1967 and 1972. The number of recorded cases continued to rise during the late seventies and began to spread out from the sugar estates to other areas of the lowveld and into parts of the adjacent middleveld.

Several factors contributed to the resurgence of malaria in the lowveld sugar estates. First, while well-maintained canals are incompatible with the reproduction of *A. gambiae,* the canals in the lowveld estates were poorly maintained and runovers and seepage occurred. This produced standing pools of water exposed to sunlight in which *A. gambiae* bred in abundance. Moreover, the estate managers did little to provide adequate housing and sanitation for their workers. Pools of stagnant water also developed in close proximity to worker housing following the rains. While spraying operations had largely eliminated the subspecies of *A. gambiae* that fed indoors, small numbers survived and were able to invade the workers' huts. Little or no effort was made by estate managers to prevent *gambiae* mosquitoes from feeding on estate workers, as huts were not screened and spraying had ceased. Renewed transmission was therefore possible.

But opportunities for transmission in the sugar estates would not have occurred without the reintroduction of malaria parasites, because the local Swazi populations were virtually free of infection, thanks to the successful earlier control program. Of the 15,692 persons tested in 1959 only 173, or 1.1 percent, were positive for malaria. The sugar industry, however, recruited a large number of migrant workers from Mozambique, where no control measures were being carried out—despite repeated warnings from health authorities that this policy could lead to a reintroduction of malaria.

The need to recruit workers from Mozambique during the mid 1950s and early 1960s resulted from the rapid influx of capital into Swaziland and the development of industries in mining, forestry, and agriculture that competed for labor. Swazi workers were able to pick and choose where they worked and largely avoided employment on the sugar estates, where wages were the same as in other industries but working and living conditions were poorer. Inadequate housing, the absence of proper sanitation facilities, insufficient rations (including no provision for family rations), irregularities in payment practices, and excessive work hours figured prominently in the reasons given by Swazi workers for their avoidance of employment in the lowveld.[4] These conditions were the product of the high capital outlays required to start up production and the sugar industry's subsequent need to reduce labor costs. Rather than improve working conditions or increase wages in order to attract Swazi labor, the sugar industry chose to recruit workers from Mozambique, where unemployment was high and workers were less particular about working conditions. Coming from malarious areas in which no control measures had been initiated, these migrants reintroduced malaria parasites into Swaziland. The colonial government, for its part, was concerned about stimulating economic development in the lowveld and chose to turn a blind eye to Mozambican recruitment.

The flow of Mozambique workers slowed in the 1970s, as an economic downturn in Swaziland forced more Swazi to accept employment in the sugar industry. By the time this change occurred, however, local pockets of indigenous infection had become established over wide areas of the lowveld. In addition, the seasonal movements of Swazi workers between the lowveld sugar estates and their middleveld homes contributed to the reintroduction of malaria into the higher regions of the country.

Thus the development of the lowveld estates created conditions that encouraged the breeding of *A. gambiae* mosquitoes, exposed workers to infection, and led to the movement of malaria parasites back into Swaziland. By doing so, colonial and postcolonial agricultural development undermined efforts to eliminate malaria from the country. Malaria remains endemic in Swaziland, occasionally exploding into regional epidemics in the wake of major climatic events, such as the cyclone that struck southern Africa in 2000.

Agricultural Colonization in Brazil

Agricultural development policies also undermined malaria control efforts in Brazil. In the 1940s Brazil registered more than 6 million cases of malaria per year, affecting approximately one-seventh of the country's population. Following the initiation of control and, later, eradication strategies during the 1950s and 1960s, the country experienced a dramatic decline in cases, reaching a low of just over 52,000 cases in 1970. By 1980, however, the annual number of cases had risen to nearly 170,000. The number of cases continued to multiply over the following two decades, reaching 630,000 by 1999. While population growth accounted for some of this rise, annual estimates of the parasite index or API (the number of positive slide cases per total slides taken) increased dramatically over this same period. The rise in malaria cases in Brazil was not evenly distributed, however, with 99 percent of registered malaria cases in the 1990s occurring in the Amazon forest regions of the country.[5]

Malaria had occurred in the Amazon as early as the seventeenth century in association with human development of the forest. However, extensive exploitation of the forest dates from the end of the nineteenth century and the growth of the rubber industry. As we saw in chapter 4, the industry attracted migrant workers from the northeastern region of the country, especially during periods of drought and famine. The pace of human incursions into the forest accelerated in the 1960s and 1970s, in response to the government's efforts to integrate the region more closely into the Brazilian economy. The government's integration policies had two goals. The first was to secure the country's borders by populating the large areas of forestlands bordering on neighboring countries. More importantly, integration was seen as a safety valve, relieving the problem of landlessness and poverty that threatened the country's social stability, by opening up new lands for agricultural settlement.

The government initiated a colonization program aimed at providing millions of landless Brazilians with opportunities to obtain cheap land. The problem of landlessness in Brazil, as in many other parts of Latin America, was not caused by an absolute absence of available land. It was instead a product of massive inequalities in landownership, with much of the country's farming and grazing land held by a relatively few wealthy landowners. Amazon colonization was an alternative to what the government perceived to be a more socially disruptive and politically unacceptable program of land reform throughout the country.[6]

The integration program began with the construction of major highways through the Amazon forest. This was followed by colonization schemes designed to encourage the creation of small and medium-size farms. Between 1970 and 1980, nearly 1 million immigrants settled in the newly developed areas, with the majority going to the state of Rondonia. Some 30,000 new farms were started. In addition, thousands of migrants poured into gold-producing areas of the Amazon, creating sprawling deforested moonscapes similar to those located along the Thai-Cambodia border, mentioned in chapter 6.

While the government initiated Amazon colonization, it did little to provide needed infrastructure to support the new agricultural communities. Health services, in particular, were rudimentary and thinly distributed. Sanitation and public health services were also largely lacking. In addition, the process of colonization created conditions that encouraged the transmission of malaria. Deforestation and settlement created breeding sites for local malaria vectors. The primary vector in the region, *A. darlingi,* bred in sheltered areas along the edges of rivers and in the pools of water that frequently developed beyond the riverbanks when rivers flooded during the rainy season. Poorly drained road causeways, burrow pits in newly created settlements, and mining sites mimicked these floodwater impoundments, multiplying the number of potential breeding sites. The gold mining areas were particularly prone to the creation of depressions where water collected. In addition, the felling of trees and clearing of farmlands created new water catchments. Soil erosion also increased the silting of rivers, encouraging the formation of embankments, which inhibited floodwaters from flowing back into rivers, further facilitating the creation of floodwater impoundments. In short, Amazon development altered the distribution and flow of water across the land in numerous ways and, in doing so, increased the breeding potential of *A. darlingi* mosquitoes.

Several other factors encouraged malaria transmission among new immigrants to the Amazon. First, malaria had been largely eliminated from much of Brazil, and very few of the immigrants possessed immunity to the disease. They were thus more likely to become very ill following infection, and more likely to infect anopheline vectors, than were the indigenous inhabitants of the Amazon who had been repeatedly exposed to malaria. Second, deforestation and settlement eliminated many of the wild animals that had been a source of blood meals for *A. darlingi*. Without this diversion, the mosquito began feeding almost exclusively on human hosts. Third, the new immigrants settled in close proximity to breeding sites and constructed shelters that provided little protection from malaria vectors. These shelters were poorly constructed, without solid walls and thus were not conducive to the use of residual pesticides. Finally, the high mobility of the migrants, their ignorance of the problem of malaria, and their lack of financial resources to protect themselves from infection or acquire antimalarial drugs contributed to the explosive growth in malaria cases.

As malaria spread through the agricultural colonies, the sickened colonists found themselves unable to plant or harvest their crops successfully. Lacking the financial reserves that would allow them to sustain even one season's losses, many migrants abandoned their plots after one or two seasons. Some sought other forms of employment in the Amazon; many flocked to the gold mining areas. But many others returned to the areas from which they had come. In doing so, they contributed to local malaria outbreaks in areas that had previously been made malaria free by control efforts. In this way, malaria expanded into areas of the country from which it had been eliminated. In 1985, 26 new foci were identified in Piauí, Ceará, Bahia, Minas Gerais, Mato Grosso do Sul, and Rio de Janeiro.

The recent histories of malaria resurgence in Swaziland and Brazil provide examples of how the burden of malaria in the post-DDT era continued to be directly linked to ongoing patterns of

agricultural development. Dozens of similar examples of development projects undermining malaria control efforts have occurred
throughout tropical regions of the globe over the past half-cen-
tury. These histories highlight the failure of government officials,
outside development agencies, and private investors to coordinate
economic development efforts with malaria control programs.
Lack of coordination and the isolation of public health from economic development planning have permitted the proliferation of
cases of so-called man-made malaria.

Patterns of economic development, however, did not always
lead to a resurgence of malaria: in 1980s El Salvador, in contrast
to the malaria experiences of Swaziland and Brazil, social and economic forces contributed to a decline, not a resurgence, in malaria.
Nonetheless, the El Salvador case points to the important roll that
social and economic forces continued to play in the epidemiology
of malaria during the post-DDT era. Malaria remained a social
disease.

Cotton in El Salvador

El Salvador is located on the Pacific coast of Central America. The
country's landscape is dominated by two mountain highlands that
transect the country from west to east. These are separated by a
wide interior valley, which opens out to the south and east into a
narrow coastal plain along the Pacific Ocean.

El Salvador joined other countries in Central America in an
effort to eradicate malaria from the Americas in 1957. At the time,
nearly the entire country was defined as malarious. Yet in most of
the highland areas of the country malaria was unstable and was in
fact easily reduced in a couple of years. By 1964 extensive residual
spraying with DDT largely eliminated malaria from areas above
650 feet. By the 1970s, the majority of cases occurred in areas
below 300 feet in the central valley and along the Pacific coast.
However, the movement of individuals between the lowlands and
highlands often triggered flare-ups in the higher altitudes, and
successive efforts during the 1970s to eliminate the remaining
pockets of malaria met with little success.

Beginning in 1980, El Salvador's Malaria Control Program,
with the assistance of advisors from the U.S. Public Health Ser-

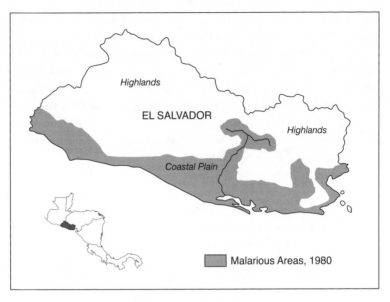

Malaria Map of El Salvador

Source: Adapted from U.S. Public Health Service, *Report of the Central America Malaria Eradication Assessment Team*, Washington, D.C., March 30, 1964, and subsequent USPHS reports.

vice, initiated an Integrated Malaria Control Program, following guidelines established by the World Health Organization. The integrated program had four elements. The first was epidemiological mapping. The country was surveyed to determine the geographical distribution of transmission. Resources for controlling malaria were then allocated to the areas with the most intense transmission, rather than equally across the whole country, as had been done in the past.

Second, the program established a system of entomological surveillance. The purpose of this system was to identify increases in vector breeding that could lead to outbreaks of transmission. The primary vector of malaria in El Salvador, *Anopheles albimanus,* preferred animals to humans; it also fed and rested out of doors, making it an inefficient vector. However, climatic conditions conducive to breeding could produce exceptionally high densities and trigger human transmission.[7] The entomological surveillance system was designed to identify any increase in vector density and initiate spraying activities before the transmission threshold was crossed.

The third element in the integrated program was environmental control aimed at permanently eliminating vector-breeding sites. Finally, and perhaps most important, was the establishment of a voluntary collaborator network. Voluntary collaborators were village health workers chosen from local communities throughout the malarious areas of the country, who were trained to identify and treat malaria cases. Persons contracting fever were directed to report their condition to their local volunteer, who obtained a blood slide and treated the patient presumptively with chloroquine. The blood slide was then sent to one of a number of labs situated throughout the malarious regions of the country to verify whether the person tested had malaria. The results of the lab tests were then quickly reported to the volunteer, who would make sure that the patient received a complete treatment of chloroquine if the test was positive.

The integrated malaria control program appears to have been highly successful in reducing malaria in El Salvador. From 1980 to 1990 malaria declined steadily from nearly 100,000 to roughly 4,500 cases per year. *Falciparum* malaria was eliminated. El Salvador was the only country in Central America to achieve this degree of success. Its success was remarkable because it occurred during El Salvador's civil war. The war not only disrupted Salvadoran society, leading to the emigration of thousands of El Salvadorans to neighboring countries, it also disrupted the economy. In the 1980s per capita income dropped by 25 percent, and the growth of GNP was a negative 1.2 percent. In addition, government resources were diverted from social and economic development programs to the military. These conditions were hardly conducive to the successful initiation of a malaria control program, especially one that depended heavily on efficient communications and stable communities within which voluntary collaborators could operate.

Those who participated in the program explained its success by describing the flexibility and commitment of the personnel involved in running the program, especially the voluntary collaborators, and this aspect of the program certainly played a role.[8] Yet other independent factors came into play.

Most of the malaria cases in 1980 were occurring in areas located along the coastal plain and central valley, a region dominated by cotton production. Cotton production had begun in the 1950s, following the initial elimination of malaria in this region using DDT. Early production was in the hands of small-scale farmers, but the completion of the littoral highway connecting the coast to the highlands in 1960 facilitated the rapid expansion of cotton production and encouraged the development of large-scale plantations owned by wealthy landlords. The area planted with cotton tripled from 106,300 acres in 1960 to 302,100 acres in the 1964–65 growing season. Cotton production continued to expand during the 1970s, sustained by the extensive use of pesticides, including DDT, to protect crops from plant-destroying insects and by the use of cheap labor. The conditions that fueled the expansion of cotton also contributed to the entrenchment of malaria along the coastal plains.

The extensive use of pesticides for agricultural purposes contributed to the development of pesticide resistance among both agricultural pests and *A. albimanus* mosquitoes. As noted in chapter 6, entomologists reported that DDT was ineffective against the primary malaria vector in El Salvador by 1958. Second-line pesticides were introduced, but the effectiveness of these was also undermined by their simultaneous use in cotton production, and resistance developed.

The employment of migrant labor also contributed to the persistence of malaria along the Pacific coastal plain. The vast majority of these workers came from highland areas where patterns of land ownership in El Salvador, stretching back to the early days of Spanish occupation, had created massive inequalities in the distribution of land. Much of the best farming land was held by a small group of landowners, and a large portion of the El Salvadoran population earned a living as tenant farmers on these large estates. Even this fragile hold on land was eliminated over wide areas of the highlands during the 1950s and 1960s, as landowners sought to maximize the productive capacity of their lands by replacing tenant farmers with modern agricultural machinery. The result was rapidly increasing landlessness. Between 1960 and 1980, the proportion of landless people grew from 12 percent of the rural population to an estimated 60 percent. It was within this context that thousands of highland residents sought seasonal employment on the lowland cotton estates.[9]

Coming from highland areas, where since the 1970s malaria had largely been brought under control, these seasonal workers had little immunity to the disease. At the same time, conditions of employment on the cotton estates exposed them to infection. Inadequate housing was especially important in the spread of malaria. Many estate owners provided no housing at all. Seasonal workers constructed open-sided multifamily shelters, called *galeras* or "chicken coops," which were often situated close to water sources, including streams and roadside drainage ditches.[10] These sites provided breeding sites for *A. albimanus* mosquitoes. Some employers provided dormitory-style housing for their workers. Yet workers often preferred to sleep out of doors because the dormitories were not well ventilated and were hot at night. The price paid for more comfortable outdoor temperatures was increased exposure to outdoor-feeding *A. albimanus.*

Finally, the system of seasonal migrant labor created temporary communities, with little social cohesion or permanency. Within these conditions the concept of a voluntary collaborator was difficult, if not impossible, to implement. Thus the conditions of production associated with cotton growing along the Pacific coast undermined efforts to eliminate malaria in El Salvador.

While the cotton industry contributed to the persistence of malaria in El Salvador, its collapse in the 1980s played a role in the success of the Integrated Malaria Control Program. In fact, the elimination of malaria in El Salvador closely paralleled the decline in the cotton industry.

Cotton production declined in El Salvador for several reasons. First, the costs of production skyrocketed. The major component in cotton production costs was pesticides. Coastal farmers were dependent on pesticide use, and as resistance developed, the costs of this dependency grew. Resistance required farmers to spray more frequently and to employ second- and third-line pesticides, which cost much more than the DDT that had initially been employed. While the cost of production increased, the price of cotton decreased as the global economic recession of the 1980s re-

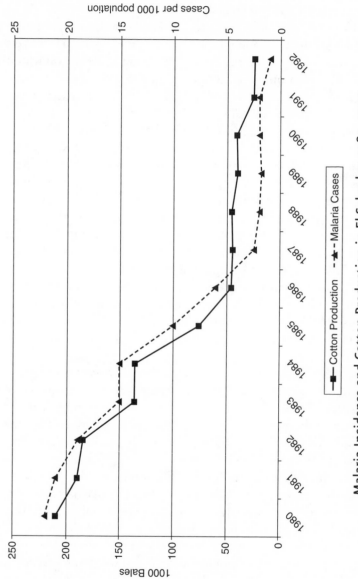

Malaria Incidence and Cotton Production in El Salvador, 1980–1990

Source: J. A. Najera et al., *Malaria Epidemics, Detection and Control Forecasting and Prevention*, Geneva, 1998 (WHO/MAL/98. 1084), 60.

Cases per 1000 population

1000 Bales

— Cotton Production   - -▲- - Malaria Cases

duced demand for cotton in textile producing countries. By 1986 the costs of production exceeded the price paid for cotton. Finally, the civil war discouraged foreign investment in El Salvador's agricultural sector. By the end of the 1980s cotton exports had fallen precipitously, from US$87 million in 1979, to US$56 million in 1983, and to only US$2.3 million in 1987.

The 1980s also saw the introduction of widespread land reform as a means of addressing the problem of landlessness, which was viewed as a major cause of civil unrest. Within the coastal plain large estates were broken up into smaller holdings and agricultural cooperatives were created.

Together, these changes eliminated the conditions that had supported malaria transmission. The flow of seasonal migrant laborers from the highlands was curtailed. Pesticide use was drastically reduced. New communities of small landholders emerged, facilitating the implementation of the volunteer collaborator network.

In the end, the success of malaria control in El Salvador during the 1980s needs to be viewed as the result of an efficiently designed malaria control program operating within a favorable social and economic environment. It is unlikely that either the program or the changing economic conditions alone would have dramatically reduced malaria. Of course, one would not want to recommend the destruction of a country's main source of export revenue as a way of controlling malaria. Nonetheless, the unintended convergence of declining cotton production with the launching of the Integrated Malaria Control Program illustrates the potential power that can be gained when malaria control strategies are supported by broader patterns of social and economic change. Conversely, the cases of Brazil and Swaziland illustrate the dangers of disconnecting malaria control from development.

WAR, REFUGEES, AND MALARIA

The global resurgence of malaria since the late 1960s has also been driven by political upheavals, warfare, and the massive displacement of human populations. Violent clashes have occurred in 21 countries in Africa since 1980 and affected more than 165 million people.[11] A map of the countries that currently experience high risk of malaria infection worldwide looks very much like the map of countries currently hosting armed conflicts. This should hardly be surprising. Malaria upsurges have been associated with nearly every major armed conflict of the nineteenth and twentieth centuries.[12]

Recent military conflicts have complex origins. Ethnic and religious divisions have played their part. However, the fragile state of newly independent governments, the resulting importance of the military in national politics, and the interference of outside governments embroiled in wider global political struggles have all fed the fires of civil war in Africa, the Middle East, Asia, and eastern Europe. So too have economic problems, which have exacerbated these ethnic and religious divisions.[13]

Warfare has increased the spread of malaria in a number of ways. It has weakened and destroyed existing health systems, making it difficult for persons infected with malaria to obtain treatment. In the Democratic Republic of the Congo, it is estimated that the ongoing civil war contributed to a 70 percent reduction in the availability of health services during the 1990s, greatly reducing access to antimalarial drugs. In some cases, warring factions directly targeted health units. Escalating warfare in Colombia in 2002 also led to the abandonment of rural health outposts. This was followed by an upsurge of cases of malaria (as well as measles). As we have seen, untreated malaria cases contribute to flare-ups, which if unchecked lead to the reestablishment of malaria in areas that had been previously free of the disease.

Not surprisingly, warfare prevented the effective operation of malaria control programs. In Nicaragua, a malaria control program using community participation methods similar to those employed in El Salvador proved successful in halving the incidence of malaria between 1980 and 1984. Yet a comparison between areas where low-intensity warfare was occurring and those where there was no combat showed a remarkably different picture. In the war zone there was a 17 percent increase in cases, while in the war-free zone there was a 67 percent decline in cases.[14] In Central Asia, malaria reemerged in Tajikistan following the break up of the Soviet Union, civil war, and the attendant collapse of preventive public health measures: the number of reported ma-

laria cases rose from 616 in 1993 to 16,521 in 1996.[15] Similarly, civil
war that began in Burundi in 1993 forced the government to cease
malaria control operations. Over the next seven years, cases went
untreated and vector breeding unchecked. Then, in 2000, abnor-
mally heavy rains led to an upsurge in vector breeding throughout
wide areas of the country. By early 2001, 3 million cases had been
reported in a population of 6.5 million people.[16]

Armed conflicts also forced people to abandon their fields and
prevented them from carrying out agricultural activities. The subsequent disruption of production in places like Somalia and the
Sudan contributed to a series of food crises in the 1990s. Extreme
malnourishment limited the ability of populations to combat the
effects of malaria infection. The disruption of agriculture also reduced incomes and the ability of individuals and families to purchase medicines to treat malaria.

Warfare also transformed local ecologies. Fields left untended
collected water, creating fresh breeding sites for malaria vectors. In
some parts of Africa, moreover, fighting has destroyed forest cover,
opening up areas to sunlight while producing shell craters and
other depressions in the soil, which collected water after the rains.
Such changes provided multiple new breeding sites for *A. gambiae* mosquitoes. During the civil war in the Congo in 1990, rebel
groups hiding out in the forest chopped down trees for firewood,
destroying large sections of the national forest reserves, contributing to an upsurge in malaria cases in the region.[17]

Warfare along with drought and famine has caused the displacement of millions of people. Forced population movements
exposed refugees to malaria. They have also introduced or reintroduced malaria parasites into formerly malaria-free areas. In the
Upper Nile region of Sudan, civil war forced tens of thousands
of people to abandon their villages and seek refuge in malarious
swamp areas in order to avoid being attacked. Malaria was the
overwhelming cause of death among Cambodian refugees who
had to pass through endemic areas on their way to camps in eastern Thailand in late 1979.[18] The movement of displaced persons
from low-lying malarious areas of Burundi during the civil war
was responsible for an expansion of malaria into the country's
highlands, where it is now endemic.[19]

Malaria is one of the primary causes of morbidity and mortality in refugee camps worldwide. For example, fever presumed to be malaria was the second leading cause of morbidity and mortality among Rwandan refugees in eastern Zaire (now the Democratic Republic of Congo) in 1994.[20] Similarly, in 1984 the annual incidence rate of malaria was 1,037 cases per thousand among Myanmar refugees in western Thailand, with over 80 percent of infections due to *Plasmodium falciparum*.[21] As with warfare, the distribution of malaria worldwide coincides with the global distribution of refugee populations. In fact, 70 percent of the people who are of concern to the United Nations High Commission for Refugees are located in malarial areas,[22] and only 4 of 43 countries with internally displaced populations are malaria free.

The number of refugees that have fled war and conflict to find shelter in neighboring countries rose from 4.6 million in 1978 to 18.2 million in 1993. There were 11.9 million refugees in 2004 (see table 2). Roughly 87 percent, or 10.4 million of these refugees were located in countries with significant malaria risks.[23]

Refugee and relief camps often created conditions that fostered the transmission of malaria, especially during the acute phase following a military conflict or the onset of famine. The presence of refugee camps, moreover, increased transmission in surrounding areas. The World Bank reported in 2003 that for every 1,000 refugees coming from a country with intense malaria, 1,400 new cases of the disease occurred in the host country.[24] The link between refugee camps and malaria and the amplifying effect that camps may have on local malaria ecologies can be seen in the history of Afghan refugees in Pakistan in the 1980s and 1990s.

Afghan Refugees in Pakistan

More than 3 million Afghan men, women, and children sought refuge in the Northwest Frontier Province of Pakistan (NWFP) following the Soviet invasion of Afghanistan in 1979. Half were able to return home in the early 1990s, after the fall of the Soviet-backed regime. But many remained, and in 2001 the refugees occupied some 200 camps located along the western border of Pakistan. The camps were set up by the Pakistan government and were supposed to house no more than 10,000 refugees each. Yet

Table 2. Refugee Populations (millions), 2004

Africa	3.2
Middle East	4.4
Southeast Asia	1.4
Other	2.9
Total	11.9

Data source: U.S. Committee for Refugees and Immigrants, Report for 2004.

a number of camps in border areas contained as many as 30,000 refugees.

Although Afghanistan had had a successful malaria control program before the war, the Soviet invasion disrupted control activities. Health expenditures were replaced by military spending, malaria control personnel were denied access to malaria prone areas, many health workers fled the country, and the Soviets are reported to have destroyed the ancient irrigation system, which was an essential part of Afghanistan's agricultural economy. In addition to reducing agricultural production, the destruction of the irrigation system allowed the formation of vast numbers of stagnant water pools, which provided breeding sites for local malaria vectors. Those who were exposed had little resistance to the disease, having been protected by the malaria control program before the Soviet invasion. Many of those who fled across the border into Pakistan carried malaria parasites with them.[25]

Conditions in the camps in Northwest Pakistan were conducive to malaria transmission and malaria quickly became a major health threat. Overcrowding, lack of adequate sanitation, and insufficient medical supplies (including bed nets and antimalarial drugs) all contributed to malaria outbreaks. In addition, a number of camps were situated in low-lying areas where water that collected following rains provided breeding sites for local malaria vectors.

An additional factor contributing to malaria transmission in the camps was the presence of large herds of cattle and goats. It is estimated that the refugees brought more that 3 million head of cattle with them. While the presence of cattle near human settlements in Europe and North America contributed to a decrease in malaria, the opposite occurred in the Pakistan refugee camps. In Europe and America the species of *Anopheles* responsible for malaria transmission either changed its feeding habits or was squeezed out by species that preferred animal blood, whereas in Pakistan the primary vectors for malaria, *A. stephensi* and *A. culicifacies,* fed off both humans and cattle. In normal circumstances this made them relatively inefficient transmitters of human malaria. However, the massive immigration of cattle into the refugee areas greatly increased the blood supply available to both vectors, significantly increasing their reproductive capacities. While the mosquitoes continued to feed indiscriminately off both humans and animals, the great increase in their numbers resulted in an increase in human transmission.[26]

By 1990 there were 150,000 cases being diagnosed each year by the combined health care services of the United Nations High Commission for Refugees, the government of Pakistan, and nongovernmental organizations. Of these cases, 30 percent had potentially lethal falciparum malaria.

The impact of the refugee camps on the wider epidemiology of malaria in the NWFP is difficult to measure precisely. The question is also complicated by the appropriate intent of some researchers who choose not to stigmatize refugee populations. Nonetheless, it cannot be denied that there was a strong correlation between the arrival of refugees and an upsurge in malaria in the NWFP where the refugee camps were located.

In the 1970s Pakistan had organized a successful malaria control program, following the failure of eradication efforts during the 1960s. After two years of house spraying with malathion, the annual parasite index (percentage of positive cases tested) for the country as a whole was reduced by 76 percent, with an 82 percent decline in *P. falciparum* infections. Within NWFP the overall decline was 51.2 percent, with a 79 percent decline in falciparum malaria. The overall API in NWFP was a stunningly low 0.38 percent.

The first sharp increase in malaria incidence occurred early in the 1980s, coinciding with a heavy influx of Afghan refugees. A second major epidemic in NWFP was recorded after the entry of the third wave of Afghan refugees. The fact that the rising incidence of malaria in NWFP was not mirrored in incidence rates for the rest of Pakistan, in contrast to the 1970s, suggests that the rise was related to the influx of refugees.[27] Within the NWFP, the rise in malaria was not restricted to the refugee populations. The number of recorded cases of falciparum malaria among the local Pakistani population rose from 3,060 cases in 1980 to 56,855 in 1992.

The influx of refugees may have contributed to a more general rise in malaria in NWFP in several ways. To begin with, the refugee camps became reservoirs for malaria parasites, increasing the overall parasite load within the districts within which they were located. Because the camps were not closed and the refugees shared cultural and ethnic affinities with local Pakistan populations, extensive contact occurred between the two populations. This allowed parasites to move easily across the landscape of northwest Pakistan. Second, the large camp populations quickly transformed the surrounding landscape. Areas around the refugee camps became barren, as trees and shrubs were cut for firewood. Grazing Afghan livestock further stripped the remaining vegetation. The denuded landscape provided few resting places for mosquitoes, increasing the likelihood of human vector contact in the camps and surrounding villages. The changed environment also facilitated water runoff and the creation of swampy low-lying areas suitable for vector breeding.[28] The repatriation of nearly half the Afghan refugees from the camps in NWFP beginning in the early 1990s did not appear to have dramatically reduced the prevalence of malaria in the district. This suggests that the environmental changes caused by the immigration of refugees had a more long-term impact on the malaria ecology of the region.

MALARIA AND HIV/AIDS

Adding to the growing burden of malaria worldwide in the 1980s was the emerging pandemic of HIV/AIDS. Many of the countries suffering from high rates of malaria were also engulfed in the AIDS pandemic; this was especially true in sub-Saharan Africa. The link between malaria and HIV/AIDS was debated during the early years of the AIDS pandemic. There was little conclusive evidence that either HIV infection affected the human susceptibility to malaria infections or the reverse. Recently, however, researchers have shown that there are clear associations between the two infections that run both ways. Women infected with both HIV-1 and malaria experienced consistently more peripheral and placental malaria, higher parasite densities, more febrile illnesses, severe anemia, and adverse birth outcomes than HIV-uninfected women. The proportional increase of malaria during pregnancy attributable to HIV was estimated to be 5.5 and 18.8 percent for populations with HIV prevalences of 10 and 40 percent, respectively. Maternal malaria was associated with a twofold higher HIV-1 viral concentration.[29] Studies have also shown that HIV infections can more generally increase susceptibility to malaria in adults. In areas in which malaria is unstable, such as highland areas of eastern and southern Africa, co-infection with HIV-1 and *P. falciparum* leads to increases in acute infections and death. Where malaria transmission is stable, and adult populations possess a degree of immunity to the disease, HIV co-infection has been associated with an increase in clinical cases, resulting from reduced levels of immunity. A recent study calculated that HIV-1 increased malaria incidence in 41 African countries by between .20 and 28 percent. The impact of HIV infection on malaria mortality ranged from .65 to 114 percent. The largest increases were in Botswana, South Africa, Zimbabwe, Namibia, and Swaziland, where HIV infection rates were highest, especially in rural areas, and malaria transmission was most unstable. Overall, the impact of HIV on malaria incidence was 1.3 percent and on malaria mortality 4.9 percent. While these percentages appear to be small, they produce large numbers when linked to the total number of cases and deaths reported annually in Africa. Thus the study estimated that HIV co-infection produced an excess of roughly 3 million cases and 65,000 deaths annually.[30]

It is important to recognize that AIDS is not an independent factor driving the resurgence of malaria in Africa and elsewhere. The AIDS epidemic has been amplified by many of the social and economic forces that have driven the resurgence of malaria. While early observers attributed the AIDS epidemic in Africa to the hypersexuality of African populations combined with particular cultural practices, it is now widely recognized that the AIDS epidemic has been shaped by the region's growing poverty, military conflicts involving the systematic rape of women, refugee situations, and the expansive system of labor migration. These forces have placed people, and particularly women, at risk of infection. While efforts to stem the tide of the AIDS epidemic through the distribution of condoms or abstinence programs have made considerable headway in some countries, few believe that the epidemic can be halted as long as the social and economic conditions driving it remain unchanged.

SILENT VIOLENCE: POVERTY AND MALARIA

The resurgence of malaria in tropical and subtropical regions of the globe has also been driven by less dramatic but equally important events. The growing impoverishment of many malarial countries is critical. This is especially true for sub-Saharan Africa, where it is estimated that more than 70 percent of the cases of malaria and 90 percent of malaria deaths worldwide occur.

The World Bank considers 40 countries to be in need of major debt relief. Of these, 32 are in sub-Saharan Africa, and nearly all of them have a serious malaria problem. In 2001 John Gallup and Jeffrey Sachs published their widely cited article on the "Economic Burden of Malaria."[31] The article showed that there was a high correlation between countries with low per capita GDP and those with a significant risk of malaria. The authors concluded that malaria retards economic growth and causes poverty. Yet there is clear evidence that the causation runs both ways and that, for individuals and nations, poverty contributes to malaria by exposing people to infection and undermining efforts to control the disease.

At an individual level, poverty has restricted the ability of people to protect themselves from infection and to effectively treat the infections they acquire. Even in the world's most malarious places, people who can afford adequate housing, employ insect repellents, wear protective clothing, and obtain effective prophylactic drugs can avoid contracting malaria (if they remember to take their pills). Those who cannot afford these protections are at risk of infection. A WHO-sponsored study of bed net use in 21

African countries between 1998 and 2001 revealed a strong positive correlation between individual economic status and the use
of insecticide-treated bed nets. The study also revealed a positive
correlation between antimalarial treatment of children younger
than five and economic status. The correlation was less robust
than that for bed nets.[32] However, the study did not allow for differences in the quality of medications employed. There is evidence
from a number of studies that the poor are more likely to use less
expensive antimalarials. In many cases this means chloroquine,
for which there is widespread parasite resistance. The poor also
dilute medications in order to make them last longer. They may
also try to stretch what medications they can afford, limiting their
efficacy. Lack of funds also leads mothers to delay taking their febrile children to health centers.[33] Even where malaria medications
are provided free, there are often other costs, such as bed charges,
registration fees, and the costs of additional supplies.

Poverty has also placed individuals at greater risk of infection.
Throughout this study we have seen how poverty has been a risk
factor for malaria, whether it was the hill town people who were
forced to seek employment on the *latifundia* of southern Italy, or
the landless poor of Brazil who have traveled to the Amazon forest for more than 100 years. More recently, poor farmers living in
the upland rain-fed rice-growing areas of Thailand engaged in a
range of economic activities to supplement their limited farm income during the 1980s. Some villagers participated in illegal log-
ging. Others entered the forest to poach wild animals or catch
frogs. Still others, as we saw in the previous chapter, sought to
earn money by digging for gemstones. Thai villagers living near
the Thai-Burma border engaged in smuggling activities and drug
running. All of these activities exposed the villagers to malarial
infections.[34]

Similarly, a crumbling economy drove thousands of people to
turn to illegal gold panning in low-lying, malaria-endemic areas
of Zimbabwe, in order to support themselves and their families
in the face of rising inflation, unemployment, and food shortages beginning in the early 1990s. The panners, known as Ma-
korokoza, exploited streams, rivers, alluvial sites, and abandoned

mine shafts. They often traveled with children who would help with various chores related to panning and maintaining their temporary households. The number of panners fluctuated with economic conditions. Following a major drought in 1991–92 there were an estimated 100,000 panners working in various parts of the country. But the number continued to grow and more recent estimates put the total at closer to 500,000. Thirty percent were women, many of whom used the meager income derived from panning, roughly US$36 per month, to supplement their family incomes. Many other panners were young men with no prospects for employment. The gold panners, who lived in fragile plastic shacks along the rivers, were particularly susceptible to malaria, as many migrated from nonmalaria or protected areas and had no immunity to the disease. They also lacked sufficient knowledge of the potentially fatal disease and took no precautions to protect themselves.[35] The plight of the gold panners contributed to a rise in Zimbabwe's malaria notification rates from around 25 per 1,000 in 1994 to 150 per 1,000 in 2002.

At a national level, resource strapped countries are often unable to support effective malaria control programs or maintain government health services. As we saw in chapter 6, lack of funds led a number of countries to cut back on their malaria control measures following the end of the Malaria Eradication Programme. At the same time, declining health budgets created shortages of effective antimalarial drugs in government health clinics. In Nigeria a rapid assessment of antimalarial drug availability found that the nonexpired supplies of chloroquine and pyrimethamine were available in only 13.3 percent of health centers, 26.1 percent of hospitals, and 27 percent at state medical stores in 2004. An average of 31 percent were available at private pharmacies and retail drug outlets. The study also showed that the distribution of antimalarial treatment guides was equally inadequate. Of 21 health centers surveyed, only 2 had an up-to-date copy of drug guidelines.[36]

Nigeria is not a unique case. A 2003 study of the availability of the nationally approved medication for malaria in government clinics in 14 African countries found that in half of the countries, the designated drug was unavailable for at least a week in the previous three-month period in at least 40 percent of the clinics surveyed. In Zambia, Tanzania, Benin, Kenya, and Uganda, 60 percent or more of the clinics reported such shortages.[37]

Availability data for chloroquine and sulphadoxine pyrimethamine (SP) tell only part of the story. With drug resistance to these two relatively inexpensive drugs rapidly spreading, their availability is less important than access to the next line of drugs. In Kenya, the government established the compound drug artemether-lumefantrine, sold under the trade name Coartem, as the country's first-line drug in 2004. A study of antimalarial drug availability subsequently found it in only 1 percent of retail outlets. More importantly, it sold for $7.60 per dose compared to $.38 for an adult dose of SP, the previous first-line drug.[38] Thus, even where the approved drugs were available, they were out of reach of the poor. As we will see in chapter 8, recent international efforts to assist African governments in procuring antimalarial medicine have helped make these drugs more available. Yet access remains a serious problem in many areas of Africa.

We need to view the links between poverty and malaria in human terms, and not just as numbers and percentages. The difficulties encountered by a poor single African mother trying to obtain malaria treatment for her son, recently captured in a television documentary, provides this human face.[39] The woman lived in the highlands of Kenya, miles from the nearest health center. Her son developed a high fever, but she had no money to purchase antimalarial treatments or to transport her son to the nearest health facility. The child languished for days before the local headman intervened, loaning her money to take the child to a local herbalist. When this did not work, she sought the assistance of one of the several untrained purveyors of biomedicine who visited the woman's village every malarial season, providing desperate parents with a combination of largely ineffective treatments. These private practitioners, locally known as "quack doctors," were unfortunately the closest semblance of Western medicine available to the villagers, without traveling considerable distances for care. Not surprisingly, none of the "quack's" treatments worked. With her child still sick and her resources exhausted, she borrowed more money from the headman to take her child to visit the nearest hospital. The child was carried by stretcher for four hours. He then had to wait two hours to catch a bus to the nearest district hospital. Once there, the mother and her sick child had to wait for hours and purchase a medical card before seeing a medical attendant. When she finally saw the attendant, she was told that although her son's antimalarial drugs would be provided free of charge, she needed to purchase a "malaria card" before her son could be treated. Charges for malaria cards and other forms of registration are ways in which government clinics exact user fees from patients in order to cover the cost of drugs and services that financially strapped governments can no longer provide for free. Unfortunately, the woman had no more money. Her son, by now critically ill, was finally treated, on credit, by the attendant, who feared he would die. After several days he was considerably better.

At this point the mother was told that she would have to pay bed charges to cover the hospital expenses before her child could be released. Again she could not pay the charges and word was sent back to her headman, who loaned her more money. Although the child was not cured, the mother removed him from the hospital, to avoid additional charges. She was given a prescription for antimalarial tablets. However, she could not afford to purchase them, and the child became sick again within a week of returning home. The woman's quest for therapy for her son had left her in debt and exhausted, and her child no better off. This all too common tale reflected the combined effects of individual and national poverty on malaria in Africa.

It is not surprising that African countries, among the poorest in the world, bear the heaviest burden of malaria. While acknowledging the importance of climate and the efficiency of local vectors, we cannot ignore the role that widespread poverty has played in the rise of malaria deaths in Africa since the mid 1960s. What may be less apparent is that many African countries have become poorer since the 1960s. In fact, Africa's growing impoverishment has mirrored the region's dramatic rise in malaria mortality. Per capita GNP for sub-Saharan Africa declined by 44.3 percent between 1960 and 1999. According to the World Bank, malaria deaths in Africa rose from roughly 100 to 160 per 100,000 over roughly this same period.[40] Africa's economic decline began with the rise in global oil prices in the 1970s and was accelerated by the global economic recession of the 1980s.

Debt and Poverty in Zambia

The entwined histories of malaria and economic development in Zambia since the 1960s provide a detailed view of the ways in which Africa's growing impoverishment has contributed to the resurgence of this disease. The case of Zambia is in some ways exceptional, because of the country's heavy reliance on copper and its support for the independence struggles of the peoples living in South Africa, Zimbabwe (at the time Rhodesia), Namibia, and the Portuguese colonies of Angola and Mozambique. Yet in many ways Zambia's experience was characteristic of events in many other sub-Saharan African countries. I have chosen to focus on Zambia because it is one of the countries that a partnership of international donor institutions has chosen to demonstrate that effective malaria control can be achieved in Africa using a series of cost-effective interventions. This effort will be examined in the final chapter of this book.

At independence in 1964, Zambia was one of the wealthiest countries in sub-Saharan Africa. The country possessed large copper reserves, which during the early 1960s fueled development efforts. The government nationalized the copper industry in order to maximize its control over export income and created a series of government-owned companies in manufacturing, trade, and agriculture. It also imposed import tariffs to encourage the development of local industries and diversify the economy. To protect farmers, the government established a guaranteed price for maize and provided subsidies for agricultural inputs. Real GDP per capita stood at around US$1,600. Per capita income was more than US$700. Both figures were among the highest in Africa. The country had a positive trade balance of more than $100 million, largely from copper exports. Copper also allowed the country to develop with limited reliance on external borrowing (in 1970, external debt stood at only $814 million). The country's leading social indicators were also positive compared to those of many other African countries.[41]

Yet the basis of Zambia's economy, copper, was also its Achilles

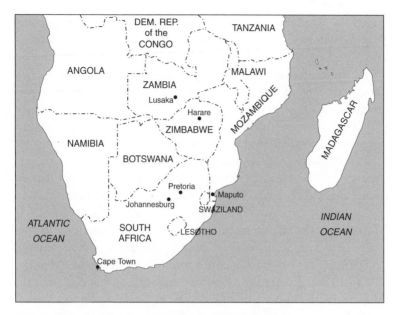

Southern Africa

heel. The economy was heavily dependent on this single resource. Agricultural exports were limited, and there was little industrial development outside the copper industry. The weakness of the agricultural sector of the economy was reflected in the high concentration of the country's population in and around the towns and cities located along the railway line that ran from the copper-producing regions in the north through Lusaka to Livingstone in the south. By the 1990s nearly 50 percent of the nation's population was urbanized.

The end of the Vietnam War in the early 1970s sparked a major drop in the worldwide demand for copper and thus in copper prices, which fell by 60 percent between 1965 and 1975. At the same time, the global oil crisis raised the cost of imports and terms of trade shifted markedly against Zambia, as they did in many other African countries during this period. Interest rates also soared.

To make matters worse, the country committed itself to ending minority white rule in southern Africa. The economic costs of Zambia's opposition to white rule were high. From the 1960s

until the late 1980s, Zambia received thousands of refugees fleeing conflict in the region. By the mid-1980s an estimated 300,000
displaced people were living within the country's borders. Zambia
lived up to its obligation to support these refugees, both morally
and materially, by diverting scarce resources to ensure their protection.

In addition, Zambia cut its trade links with the white regimes
in order to participate in international economic sanctions against
them. New trade routes had to be built so that imports and exports could be rerouted. Initially oil imports and copper exports
were airlifted through Tanzania at great expense. Later Zambia
built a road, railway, and oil pipeline to distant Dar es Salaam in
Tanzania to avoid transporting goods through Rhodesia to Mozambican and South African ports. All told, Zambia's political
commitment to liberation cost it millions in lost trade revenues.

With rising costs and declining revenues, and hoping that the
downturn in copper prices would be temporary, Zambia under-
took huge external borrowing in the 1970s in order to maintain
previous levels of spending rather than restructure its economy.
About half the debt was borrowed from multilateral funders,
chiefly the World Bank's International Development Association,
the International Monetary Fund, and the African Development
Bank. The remaining debt was owed to bilateral creditors, of
which Germany and Japan were the largest. By 1981 the country's
external debt had more than doubled, to more than US$2 bil-
lion. Zambia was not alone in this regard. Between 1970 and 1976
Africa's public debt level quadrupled. Total debt as percentage of
export goods and services rose from 10.9 percent in 1980 to 20.5
percent in 1990, and averaged 27 percent in the 1990s. As a result increasing amounts of the country's trade revenues were being
used to repay loans rather than generating further development or
paying for existing services, including health services.

In the face of rising debt payments and a stagnating economy,
the Zambian government entered into debt negotiations with the
International Monetary Fund and World Bank. The World Bank
imposed a series of structural adjustment policies designed to sta-
bilize the country's economy: liberalizing the economy; removing
import subsidies, raising interest rates; devaluing currency; cut-

ting government expenditures, including subsidies on food and fertilizer; and freezing wages.

The structural adjustment policies did not produce their intended results. Between 1983 and 1986, growth in GDP was zero, and the budget and trade deficits widened. Worse, stabilization policies, as in many other countries, led to social and political instability. Student riots against the reforms in 1984 were followed by widespread industrial unrest in 1985. Food riots broke out along the Copperbelt in 1986. The deteriorating political situation led the government to unilaterally cancel the stabilization agreement in 1987 and attempt to establish its own economic reform program. This led to a cessation of international lending and a run-up of arrear debt payments of US$3 billion. In 1989 Zambia and the World Bank entered into a new set of agreements.

The elections in 1991 brought to power a new president committed to economic reform based on the World Bank's conditions. The new regime increased efforts to liberalize trade and privatize industry. Despite the implementation of these policies, Zambia's real GNP per capita continued to drop as did per capita income. Employment also plunged following the breakup of the state-run companies and the move toward trade liberalization. Formal manufacturing employment fell from 75,400 in 1991 to 43,320 in 1998. Paid employment in mining and manufacturing fell from 140,000 in 1991 to 83,000 in 2000. Paid employment in agriculture fell from 78,000 in 1990 to 50,000 in 2000. Employment in the textile industry, one of the success stories of the country's earlier efforts at import substitution, fell from 34,000 in the early 1990s to 4,000 in 2001, as trade liberalization opened the country to a flood of secondhand clothing from Europe.[42]

Liberalization also led to the restructuring of Zambia's cotton industry. The state set up a cotton marketing board to purchase cotton and provide subsidies and agricultural inputs, such as fertilizer and seeds, to farmers in the late 1970s in hopes of increasing cotton production and diversifying the economy. Under market reform policies established in the 1990s, cotton marketing was privatized and subsidies were eliminated. Large multinational lint companies replaced the marketing board and set up "outgrower" schemes, in which small farmers received agricultural inputs on credit to be paid back at harvest time. Competition among lint companies was supposed to increase the prices paid to farmers. During the initial phase of the new program, however, the lint companies divided the country geographically and thereby avoided competing with one another for the farmers' cotton. In the face of continued low prices, many farmers dropped out of cotton production.

By the late 1990s, more cotton-buying firms had entered the competition. Yet Zambian cotton farmers benefited little from the change, as world cotton prices decreased steadily after 1995 and the costs of inputs increased. Many farmers found themselves unable to meet their debts and, like sharecroppers in the American South after the Civil War, entered into a kind of debt-peonage relationship with the lint companies. A number of factors led to declining world cotton prices; however, a major contributor was the subsidies that the U.S. government paid to U.S. cotton farmers. These subsidies distorted international markets by encouraging U.S. farmers to continue producing at high levels despite falling prices. This surplus production, much of which was exported, suppressed international cotton prices. The International Cotton Advisory Committee estimated that the removal of these subsidies would result in a 10 percent reduction in U.S. cotton production, and a $.11 or 26 percent increase in world cotton prices. Oxfam calculated that sub-Saharan exporters lost $US302 million as a direct consequence of U.S. cotton subsidies in 2001. For Zambia, this meant a loss of US$8 million in export earnings in 2001–2, which equals roughly 1 percent of Zambia's GDP and total export earnings.[43] A report by two World Bank economists calculated that the removal of U.S. cotton subsidies would increase the per capita income of Zambian cotton farmers by roughly 1 percent. While this is a relatively modest amount, they noted that market participation in cotton was highly elastic, and that the higher prices would stimulate more farmers to engage in cotton production. They calculated that this could lead to an increase of 24 percent in the household incomes of poor rural families who devoted 1.2 hectares of land to cotton.[44] For this to happen, however, there would need to be a system of price supports that would protect farmers from drastic fluctuations in international cotton prices.

Such supports, available to U.S. producers, were removed in Zambia in response to World Bank specifications aimed at reducing government expenditures and liberalizing trade. U.S. cotton subsidies, combined with the Bank's restrictions on subsidies in Zambia, reduced government resources, limited efforts to diversify the economy, and discouraged poor rural families from engaging in an important income-generating activity.

Zambia's indebtedness also increased in the wake of liberalization. The ratio of debt to gross domestic product rose 200 percent in the 1990s. By 2002 total debt had risen to more than US$6 billion and debt service payments had risen to more than US$600 million per year. In December 2000 Zambia became eligible for debt relief under the World Bank's Heavily Indebted Poor Countries program. The program granted limited amounts of debt relief to countries that had met a set of specific economic targets under the Bank's structural adjustment programs. However, this intermediate debt relief covered only 5 percent of Zambia's debt.[45] By 2003 per capita debt stood at $700, while per capita income had fallen to less than $1 a day ($358), or roughly half of what it had been at independence. According to the Zambian Government's Living Condition Monitoring Survey, nearly 70 percent of the population fell below the poverty line (measured as per capita monthly income below US$3.05) in 2002–3.

Zambia's debt problems had a devastating effect on its abilities to provide for the social needs of its population and particularly the poor. Total per government capita expenditure on health in 2005 was US$11, a figure that placed the country 152 out of 186 countries worldwide. During the 1990s government-run health services throughout the country deteriorated. Bank-mandated conditionalities required Zambia to establish a wage cap in order to reduce overall expenditures. Low wages and disruptions in pay caused many health workers to opt out of the health system. Skilled workers sought better-paying jobs in health systems in South Africa or abroad (in 1993, 3,444 out of a total of 9,000 trained nurses applied to work abroad).[46]

Declining numbers of health workers caused the government to shut down health facilities. In the rural areas in particular, basic supplies, including antimalarial drugs and medical services, be-

came widely unavailable. The accessibility of health care deteriorated, particularly in remote areas. In Western Province in the late
1990s, the nearest health clinic to the Luampungu villages was 28
miles away. Travel was difficult in the sandy soil and patients had
to be transported by ox-drawn sledge for several days to reach the
clinic. If the clinic did not have the right drugs, as was often the
case, patients needed to proceed to the mission hospital, another
25 miles away. The WHO estimated in 2005 that only half of all
rural households were located within 3 miles of a health facility.

For the average Zambian in the 1990s, health services not only
declined in quality but also increased in cost, as external pressures for health sector reform forced the government to implement user fees for health services, austerity measures, and priva-
tization. User fees introduced in 1993 compounded the effects of
declining incomes and increased unemployment making health
care inaccessible to a growing proportion of Zambians. The next
year, the World Bank reported that following the imposition of
user fees in Lusaka visits to outpatient clinics fell by 60 percent
and to maternal delivery services by 20 percent.[47] Attempts to provide subsidized care for the most vulnerable populations through
a Community Health Waiver Scheme did little to offset the negative impact of fees on clinic attendance in rural areas.[48]

The global AIDS pandemic further stretched the country's
declining health resources. Nearly one in five people in Zambia
was infected with HIV in 2005. The ratio was higher in the urban areas. The AIDS epidemic contributed to a sharp decline in
life expectancy from 54.2 years in 1990 to 33.4 years in 2001. The
country made a strong political commitment to coping with the
AIDS crisis. However, efforts to do so placed an additional burden
on the already weakened health system. An estimated 200,000
people were in need of treatment in 2005.[49]

Not surprisingly, Zambia's malaria control program suffered
from the country's declining economic resources. Malaria control
at the end of World War II focused primarily on controlling the
disease in the urban centers and mining areas of the country by
treating houses with DDT. Malaria was almost eliminated in these
areas by 1970, though it remained a serious problem in the rural
areas, where no effective control measures were being applied. In

the wake of the economic downturn in the 1970s, malaria control measures began to break down in the urban areas. Local authorities, responsible for eradication, faced critical financial constraints that led to the discontinuation of a number of social services. The spraying program was forced to redeploy spraymen or lay them off. A WHO delegation that visited regular spraying sites in 1973 found that up to 30 percent of them had not been covered. The widespread international opposition to the use of DDT forced Zambia, like many other countries, to shift to more expensive pesticides. In Zambia malathion became the first-line pesticide. Malathion costs three to five times as much as DDT, and by the end of the 1970s only the mining industry could afford to use the pesticide to protect their workers. By the early 1980s, falling copper prices and rising oil prices forced the mines to reduce their spraying operations. They began by withdrawing spraying from buffer zones—the residential areas surrounding the mining compounds—and then in 1990 ceased residual-spraying operations altogether.[50]

Declines in health funding also caused antimalarial drug shortages in government-run clinics and hospitals. Drug outages, combined with the inability of a growing proportion of the population to afford drugs purchased from private sources, contributed to intermittent patterns of chloroquine usage. This led, in turn, to growing chloroquine resistance, ranging from 12.8 percent in the Western Province to more than 50 percent in the Southern Province and Copperbelt by 2000.[51] Resistance to SP also began to emerge in the 1990s. This rising drug resistance contributed to a sharp rise in death rates in patients admitted to hospitals with the disease between 1985 and 1994. The situation worsened in 2001, when the government switched to the combination therapy Coartem as the front-line drug against malaria: its relatively high price compared to the cost of earlier first-line antimalarials limited access to the drug.[52] Failure to fully cure patients infected with malaria, whether because of drug resistance or a lack of access to drugs, resulted in growing numbers of human gametocyte carriers who contributed to the spread of the disease.

Economic conditions also drove the resurgence of malaria in Zambia by exposing an increasing number of people to infection.

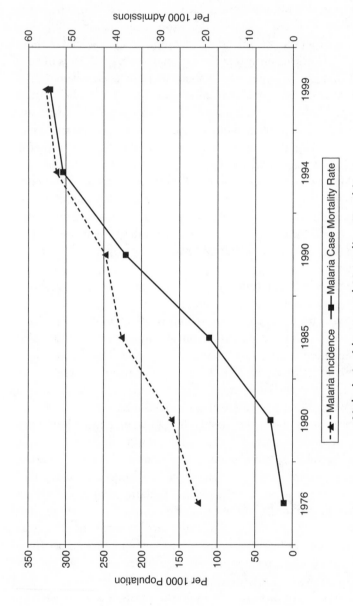

Malaria Incidence and Mortality, Zambia

- ★ - Malaria Incidence   ─■─ Malaria Case Mortality Rate

Source of data: *Malaria in Zambia. Situation Analysis, May 2000.* National Malaria Control Centre, Central Board of Health (Lusaka, Zambia, 2000).

This was particularly so in urban areas that had formerly been protected. With price guarantees and subsidies removed, rural farmers unable to maintain production sought employment in the cities. The size of the urban population grew as these new urban settlers moved into sprawling informal settlements, which lacked adequate sanitation, water, and drainage. With the breakdown in spraying activities, these settlements became breeding grounds for *A. gambiae* mosquitoes and a focus of increased urban transmission of malaria. The overall well-being of urban residents along the Copperbelt was undermined by World Bank–mandated economic policies, which acted to depress wages at the same time that they increased the price of imported goods. By the mid-1990s, the country's continued economic decline had begun to reverse the process of urbanization as urban job opportunities vanished, even in the so-called informal sector. Urban residents began moving back to the countryside to eke out a subsistence living on the land.[53] This reverse migration resulted in the movement of people with limited exposure to malaria into highly endemic parts of the country, where malaria had never been brought under control.

Overall the annual incidence of reported of malaria cases nationwide rose steadily from 121 per 1,000 in 1976 to 376 per 1,000 in 2000.[54] Worse yet, the incidence in children younger than five was more than 900 per 1,000. By 2003 the disease was endemic to varying extents in all nine provinces of the country: in the low-lying areas, malaria was hyperendemic, with infant infection rates of more than 50 percent; in the highlands, infant infection rates were less than 10 percent. The heavily populated districts along the main railway line were subject to periodic malaria epidemics. Countrywide, malaria accounted for 37 percent of all outpatient hospital visits, 62 percent of all hospital admissions, and 40 percent of all infant deaths; 95 percent of all cases were *P. falciparum* malaria. In all of Zambia, malaria accounted for 50,000 deaths a year.[55]

❖ ❖ ❖

Since the late 1960s malaria has been on the rise across wide areas of the globe, regaining much of the ground it had lost during the DDT era in large areas of southern and southeastern Asia, as well as in much of Latin America. While the rise in malaria cases and deaths did not come close to pre-eradication levels, and some of the initial increases were brought down again in the 1990s, malaria remained a significant problem in these areas. In most of sub-Saharan Africa, malaria had never been brought under control. Yet Africa too experienced an upsurge of cases and deaths between 1970 and 2000. Moreover, there was little indication that the situation had improved by the end of the century. Only in Zimbabwe, Swaziland, and South Africa had significant progress been made in reducing malaria prior to 1970, yet malaria returned across southern Africa in the 1970s and 1980s.

The global resurgence of malaria was driven by a combination of malaria control program failures, the declining efficacy of first-line pesticides and antimalarial drugs, misguided development policies, armed conflicts, population displacements, the AIDS pandemic, growing debt burdens, collapsing health systems, and rising poverty. While many of the conditions driving the growth of malaria were new, the role of social forces in the epidemiology of this disease was not. Malaria continued to be a social disease. Moreover, the distribution of these forces, while widespread, was concentrated with particular intensity within the tropical south and particularly in sub-Saharan Africa, contributing further to the evolution of malaria as a tropical disease.

Against this array of forces, the world public health community in 1998 launched a new attack aimed at dramatically reversing the global resurgence of malaria, particularly in Africa. The new attack, operating under the slogan "Roll Back Malaria," had the ambitious goal of reducing the global burden of malaria 50 percent by 2010 and 75 percent by 2015. The attack was centered on the application of a series of specific interventions, which had been shown to be effective in reducing malaria-related mortality in children but which largely ignored the forces that have driven the resurgence of malaria in places like Zambia over the past 40 years. It is perhaps fitting then that the success or failure of this new attack appears likely to be played out in this impoverished Central African country. It is to the history of Roll Back Malaria, and its prospects for success in Zambia and elsewhere, that we turn in the final chapter of this book.

CHAPTER EIGHT

Rolling Back Malaria
The Future of a Tropical Disease?

In 1992 ministers of health from 102 countries convened in Amsterdam to discuss ways of responding to the global resurgence of malaria. The conference was convened to raise public awareness of the disease and stimulate national and international actions to curb it. The conference concluded by endorsing a new Global Strategy for Malaria Control.

The new strategy reflected lessons that had been learned from past control efforts, particularly the failed Malaria Eradication Programme (MEP). It also embodied the principles of selective primary health care that had emerged in the 1980s. Selective primary health care centered on the provision of a series of narrowly focused cost effective interventions aimed primarily at reducing infant and child mortality. The Global Strategy for Malaria Control emphasized the need to shift from highly prescriptive, centralized programs to flexible, cost-effective, and sustainable programs adapted to local conditions and responsive to local needs.[1] The new malaria control efforts were to be integrated into existing health services and were not intended to act as stand-alone programs. An equally important part of the new Global Strategy was the call for a multisector approach to malaria control. The strategy paper noted, "Malaria control is not the isolated concern of the health worker. It requires partnership of community members and the involvement of those involved in education, the environment in general, and water supply, sanitation and community development in particular. Malaria control must be an integral part of national health development and health concerns must be an integral part of national development programmes."[2]

The principles and approaches embodied in the Global Strategy for Malaria Control were subsequently integrated into a new multilateral program to Roll Back Malaria (RBM), which was proposed in 1998 by the incoming World Health Organization (WHO) director general, Dr. Gro Harlem Brundtland and jointly sponsored by the United Nations Children's Fund (UNICEF), the World Bank, and the United Nations Development Programme (UNDP). Roll Back Malaria was intended to be a partnership of malaria-affected countries, UN agencies, development banks, bilateral donors, Organization of Economic Cooperation and Development donor countries, the research and control communities, industry, the private sector, and nongovernmental organizations. The RBM Partnership was created to ensure that countries had effective access to the information, technology, and financial resources required to reduce their burden of malaria.

Like the Global Strategy, the RBM Partnership emphasized the need to employ cost effective sustainable approaches to malaria control. Of particular importance in highly endemic areas was the use of insecticide-treated bed nets (ITNs). Bed nets had been used since antiquity to protect sleeping individuals from biting mosquitoes. Herodotus, writing in the fifth century BCE, described how Egyptians living in marshy lowlands protected themselves with fishing nets. By the early nineteenth century, British colonists in India—most likely inspired by the example of Punjabi fishermen—were also sleeping under nets. However, it was not until World War II that nets and insecticides were combined. The American military in the Pacific theater impregnated bed nets and jungle hammocks with 5 percent DDT to ward off malaria and filariasis. The use of insecticide-treated nets using synthetic pyrethroids emerged as an alternative to residual house spraying in the 1970s.[3] ITNs repelled and killed more mosquitoes than simple nets and were shown to reduce child deaths by 16 to 33 percent in studies conducted in the 1990s. Outcomes were best where transmission levels were lowest, yet even in very high transmission areas, such as in western Kenya, the nets resulted in a 90 percent reduction in malaria transmission and made a significant difference in mortality.

Roll Back Malaria also endorsed the use of intermittent preventive therapy with antimalarial drugs to protect women from malaria during pregnancy. Women who become infected with malaria during pregnancy experience high levels of maternal anemia. This contributes to the birth of low-weight babies, who in turn are at increased risk of premature death. In low transmission areas, malaria in pregnancy among women who have no immunity to malaria may also cause miscarriages. In sub-Saharan Africa, malaria infection is estimated to cause 400,000 cases of severe maternal anemia and from 75,000 to 200,000 infant deaths annually. Maternal anemia contributes significantly to maternal mortality, causing an estimated 10,000 deaths per year.[4]

A third approach of Roll Back Malaria was the early detection and treatment of children with fever. Delays in treatment are a major contributor to child mortality from malaria. RBM guidelines recommended that children with fever should be diagnosed and treated within 24 hours of the initial onset of symptoms.

RBM also called for the early detection and prevention of malaria epidemics in areas of "unstable" malaria. As we have seen in previous chapters, the invasion of falciparum malaria into normally nonmalarious regions often resulted in widespread killing epidemics. The RBM guidelines called for the identification of an epidemic situation within two weeks of its onset, and the provision of effective control measures within two weeks of notification.

Finally, following the Global Strategy, RBM endorsed a multisector approach, involving a range of government ministries, and called for the reinforcement of basic health services. Its goal was to reduce the global burden of malaria by half by 2010 and by 75 percent by 2015.[5]

Although the goal of the RBM Partnership was to reduce the global burden of malaria, its initial and primary focus was on reducing the burden of malaria in Africa. Africa has been a particular challenge to malaria control. While malaria control efforts had had some successes in other parts of the world, little has been achieved in sub-Saharan Africa outside of southern Africa. In addition, previous strategies, which relied on indoor residual spraying, were believed to be ineffective in most of the region. A study in Garki, Nigeria, in the 1970s found that even a perfectly designed spraying program, which eliminated more than 90 percent of malarial vectors, did not curtail transmission.[6] The RBM interventions were extensively tested in Africa and were believed to provide a way forward. Delegates to a 2000 conference in Abuja, Nigeria—convened to kick off Roll Back Malaria in Africa—committed themselves to adopting the proposed interventions and set an ambitious goal of attaining 60 percent coverage with the interventions by 2005.

Faith in the ability of the partnership to achieve its goals rested on the belief that its strategy represented a new approach to malaria control that would overcome the problems encountered by earlier efforts, and particularly by the WHO's ill-fated Malaria Eradication Programme of the 1950s and 1960s. But did RBM represent a new approach? Or was RBM simply old wine in a new bottle? Was it one more effort to mobilize flagging support for malaria control by promising to achieve definable goals? Was it one more effort to control malaria by applying simple technologies, while largely ignoring the underlying forces driving malaria transmission?

A NEW APPROACH TO MALARIA CONTROL?

In a number of ways, Roll Back Malaria did mark a clear departure from the earlier Malaria Eradication Programme (MEP). The goals of the RBM Partnership were more limited than those of the MEP, recognizing both the risks and difficulties of promising the total elimination of malaria.

The MEP had been largely directed by WHO, with logistical support provided by UNICEF and bilateral donors. The RBM Partnership involved a wider constellation of organizations. The most important of the new participants was the World Bank. The Bank had not been involved in funding health programs at the time the MEP was initiated in the 1950s and 1960s, but entered the field in the 1970s with the Bank's investments in river blindness programs in West Africa. In the early 1990s, the Bank's World Development Report, *Investing in Health,* announced its commitment to scaling up investments in health as a way of generating global economic development. The participation of the World Bank was crucial to the success of RBM, as it had the ability to direct massive amounts of money toward malaria control. The Bank could also leverage country governments to devote resources to malaria. In the Zambian case, for example, the government included malaria control in the Poverty Reduction Strategy, which the country was required to follow in order to maintain Bank support and become eligible for debt relief.

RBM also created public-private partnerships designed to expand the resource base for malaria control—for example, it persuaded the giant drug company Novartis to provide at cost the new artemisinin combination drug Coartem.[7]

Unlike the MEP, Roll Back Malaria was not organized around a single intervention applied in a "one size fits all" manner but employed a range of methods to control malaria, each adapted to local circumstances. In a change from most postwar programs, these interventions were also not directed at vector control. While bed nets were often talked about in program documents as a vector control method, the ways in which they were deployed under RBM, at least initially, were not designed to significantly reduce the mosquito population. In highly endemic areas, ITNs could only achieve an effective level of vector control if at least 70 percent of the entire population were protected. RBM's approach was focused on employing nets to protect infants and pregnant women.[8]

The RBM interventions were also different in that they required people to change their behavior. In order for RBM to be successful, individuals had to acquire and to properly use and maintain treated bed nets. Pregnant women needed to seek antenatal care and follow preventive regimens. Families had to learn to recognize the symptoms of malaria and quickly obtain medical care when a child developed a fever.

The funding of Roll Back Malaria interventions reflected a new emphasis on individual responsibility for health care and for the financing of malaria control measures. During the early years of the program, bed nets were not distributed like vaccinations or house spraying was, but had to be purchased. Health services often required the payment of user fees. This emphasis on self-financing reflected the World Bank's wider approach to health sector reform implemented in the 1980s and 1990s.

Finally, RBM was also committed, at least on paper, to pursuing a multisectoral approach that involved improvements in education, the water supply, sanitation, environment, and community development. It also recognized the importance of supporting the development of local health services and making its strategies part of a sectorwide approach to promoting health.

In other important ways, however, RBM was similar to the MEP. Despite its commitment to self-financing models, the program's strategies could not be implemented without the provision of substantial subsidies. Although local governments were expected to provide a large portion of these payments, the success of RBM was dependent on large amounts of external funding. As with the MEP, the funds available were not sufficient to meet the partnership's goals during its first five years. Moreover, the estimates of what it would take to achieve its goals increased during this initial period, as they had with the MEP. This increase was largely caused by the development of drug resistance and the need for many countries to switch over to combination therapies.

The use of combination therapies was modeled on earlier efforts to restrict the development of drug-resistant tuberculosis. Attacking the malaria parasites with multiple drugs reduced the possibility that resistant strains would survive and multiply. Combination drug therapy also reduced the treatment time from seven days to three days, which reduced parasite exposure to each drug and thus the risk of resistance developing. A number of new combinations included the compound artemisinin, which was shown to be highly effective against *P. falciparum* malaria. The cost of such combination therapies, and especially those which included artemisinin, was considerably higher than that of simple therapies such as chloroquine and sulphadoxine pyrimethamine (SP). For example, the cost of Coartem, one of the artemisinin-based combination therapies, currently provided to African countries by Novartis "at cost," is $2.40 per adult course. This compares with $.10 retail for chloroquine.[9]

Prior to 2003, moneys to support RBM initiatives were slow to reach national malaria control programs. In order to obtain funding, malaria-infected countries were required to develop country-wide control plans following RBM guidelines. By 2001, 13 African countries had developed national strategies. The combined projected budgets required to complete the first year of these plans was just under US$150 million. At the time, however, only US$32 million had been pledged, or 21 percent of what was needed. By the end of 2004, nearly US$6 billion had been pledged by donors, but only US$146 million had been disbursed. The World Bank pledged $350 million to $500 million for RBM at the Abuja conference. However, in distinct contrast to the Bank's early contributions to fighting AIDS, this pledge was not new money. Instead, it represented commitments that individual countries were expected to make to RBM initiatives from the International Development Association credits they received from the Bank in three-year cycles. Individual countries, however, were hesitant to devote these credits to malaria because these credits were normally used to fund portions of development budgets for which there were no other possible donors. Health, on the other hand, was widely viewed as an area for which there were potential external donors. In fact, up to 80 percent of the health budgets of some African countries were funded from external sources. The Bank for its part did not take an active role in encouraging countries to devote credits to malaria until 2004.[10]

The slow trickling of funds to national programs meant that the roll out of interventions was delayed. Stakeholder frustration over the failure of global funders to provide promised funds in a timely manner, once national plans were in place, was evident in the responses that national health authorities in several countries gave to RBM's external evaluators in 2002. A representative from Malawi complained:

> We've done everything we've been asked to by RBM. We've had donor meetings, we've produced documents, but we haven't seen the funding. . . . It's taken two years to develop the plan and budget, but no donor has said that they are ready to work with

the initiative . . . where's the money? There was a huge drive by government to finish the document and put a budget on it. The budgeted plan is a white elephant.[11]

And like the MEP, the RBM partners failed to adequately address the need to invest directly in strengthening the health systems of participating countries, particularly in Africa. Despite commitments to health-sector development in the original RBM mission statements, and the role that poorly developed health services played in restricting earlier efforts to control malaria, the RBM partners focused on the distribution of materials needed for malaria control rather then on addressing the underlying weaknesses of health systems. Lack of trained staff, gaps in health service coverage, weaknesses in the administrative capacity of health systems—particularly in the area of procurement—and the lack of effective monitoring and surveillance capacities all limited the ability of RBM partners to meet their goals during the first five years of the program.

RBM also resembled MEP in that it failed to seriously address the underlying forces driving transmission. On paper, RBM viewed malaria as a multisectoral problem, involving many areas of social and economic development. Yet little was done either to coordinate malaria control with broader development efforts or to adjust control strategies to local economic and social realities. The United Nations Development Programme, which was one of the primary partners in RBM, was supposed to be responsible for assisting countries in this regard. Yet until 2005 it was largely absent from the RBM Partnership. As a representative from Zambia noted, "UNDP is invisible in Zambia. They don't talk to their Head Office. Their Head Office knows about RBM—but the Country Office has no idea . . . they don't believe that malaria is in their portfolio. Why hasn't Head Office informed the Country Office?"[12] Without this type of coordination, RBM interventions were unlikely to be supported by parallel improvements in social and economic conditions.

Another similarity between the RBM and MEP is that both were promoted as contributing to economic development. This replaying of old economic arguments occurred despite the inabil-

ity of MEP supporters to demonstrate that malaria eradication
had contributed to economic development and the presence of a
substantial body of writing by economists that was critical of the
methods employed to make a link between malaria control and
development. Moreover, a number of studies in the 1980s and
1990s showed that in holoendemic areas of Africa, malaria control
had little impact on the economy.[13]

RBM's claims regarding the economic benefits of malaria control, captured in the phrase, "Roll Back Malaria, Roll in Development," were largely based on the writings of economist Jeffrey
Sachs and his colleagues at Harvard and the London School of
Tropical Medicine and Hygiene. In particular, his 2001 article in
the *American Journal of Tropical Medicine and Hygiene* was widely
cited by Roll Back Malaria leaders as evidence that malaria contributed to economic underdevelopment and that the reduction
of malaria would contribute to economic development.[14] The article was so influential that visitors to the RBM Web site, as well
as the Web sites of the WHO and World Bank, were invited to
download this article as well as other writings by Sachs.

Gallup and Sachs concluded that per capita GDP in economies with malaria grew 1.3 percent less per year from 1965 to 1990
than those without malaria and that a 10 percent reduction of
malaria was associated with a 0.3 percent higher growth in per
capita GDP. They based their conclusions on a multiple regression analysis of the effect of eight variables, including malaria, on
per capita GDP in 150 countries. The analysis showed that there
was a strong correlation between malaria and decreased per capita
GDP, even when a range of other causal variables—"distance to
markets," "presence of hydrocarbons," "tropical land areas," and
"trade openness" were entered into the analysis.

Unfortunately, a close reading of Gallup and Sachs raises serious questions about both the methodology they employed to
achieve their findings and the conclusions they reached. Most
importantly, their analysis only allowed them to demonstrate
that there was a correlation between the presence of malaria and
decreased per capita GDP. It did not permit them to determine
which way the correlation ran. This is because regression analyses only tell you how strong an association there is between two

variables. They do not tell you how the two variables are related. Gallup and Sachs avoided this issue by somewhat disingenuously denying that poverty could contribute to malaria. However, in a subsequent article in *Nature,* Sachs did acknowledge that:

> It is certainly true that poverty itself can be held accountable for some of the intense malaria transmission recorded in the poorest countries. Personal expenditures on prevention methods such as bed nets or insecticides, increased funding for government control programmes and general development such as increased urbanization can reduce malaria transmission.[15]

As we have seen throughout this book, malaria and poverty are mutually reinforcing. This fact alone makes it difficult to place precise percentages on the impact of malaria on economic development on the basis of a regression analysis.[16] The study's findings are further thrown into question by the variables the authors chose to measure against malaria, or rather by the variables they did not include. For example, no economist would deny that Zambia's massive external debt and debt service payments have restricted its economic growth during the 1970s, 1980s, and 1990s. Yet, remarkably, Sachs and his coauthors did not include levels of external debt and debt service in their list of variables that might affect a country's economic performance. Nor did they include a country's degree of economic diversification. Yet the dependence of countries like Zambia on one or two export items for revenue has clearly made them vulnerable to international price fluctuations and restricted their economic growth.

RBM resembled MEP in one final way. Although RBM stressed the application of multiple strategies adjusted to fit local settings, there was relatively little room for adjusting the mix of interventions that could be employed by national malaria control programs. Countries could determine how they would distribute nets;[17] but they found it difficult to reject the use of ITNs or to employ an intervention that was not part of the RBM's technical components. In particular, until 2006 RBM partners discouraged the use of indoor residual spraying with pesticides to control malaria in endemic countries in Africa. This position was fueled by the failure of the MEP, the results of the Garki Project noted earlier, and WHO's movement away from centralized vertical malaria control programs and toward the integration of control into community-level health services during the late 1970s and 1980s. In 1985 the World Health Assembly resolution 38.24 called on malaria-endemic countries to move away from spraying and to dismantle the vertical malaria control programs and move towards a more horizontal "community-based" system of malaria control.

Cost-benefit analyses in several countries in which malaria was highly endemic also suggested that ITNs were a more cost-effective weapon than indoor residual spraying in controlling malaria. Nonetheless, in areas of unstable malaria such as Ethiopia, Zimbabwe, and South Africa, where indoor residual spraying had been successfully used to control malaria in the past, RBM met with considerable resistance from local health authorities who wished to continue employing this strategy and needed external financial support to meet the rising costs of pesticides as well as to train personnel to run spraying programs (in 2005 malaria control workers in Ethiopia were still employing malaria maps that had been produced in the 1950s). RBM partners, following WHO technical guidelines, declined to provide funds to support these activities, arguing that spraying was not cost effective compared to the use of ITNs. The only exception to this was in areas of complex emergencies, such as major epidemics caused by the invasion of malaria into highlands areas, and in the World Bank–supported projects in Eritrea and Mozambique. Since 2005, the partnership has lessened its resistance to spraying strategies, though the main focus of its interventions aimed at reducing transmission in most of Africa remains centered on the use of nets.[18]

EVALUATING RBM, 1998–2004

Despite its flaws, RBM was successful in a number of ways during its first five years. In addition to greatly raising awareness of the problem of malaria, it succeeded in bringing much greater levels of funding to bear on the problem than had existed since the end of the MEP, even though the money was often slow to funnel down into national programs. By raising awareness of the global burden of malaria, the program also helped to develop other initiatives aimed at fighting malaria. The Global Fund for AIDS,

Tuberculosis and Malaria, established in 2002, had by 2006 approved more than US$2 billion to support malaria control programs, primarily through the procurement of antimalarial drugs and ITNs. These funds helped kick start the rapid deployment of RBM interventions. The Gates Foundation also committed US$268 million, primarily for the development of vaccines, new drugs, and pesticides. In July 2005 the Bush administration committed US$1.2 billion for combating malaria over five years.

Vaccine Development

Investments in vaccine development also increased dramatically as a result of RBM. Questions concerning the financial viability of malaria vaccine research for a long time limited efforts to develop a vaccine. After all, most of those who needed the vaccine had little ability to pay for it. In addition, research that was funded with public dollars carried with it restrictions on how profits from the vaccine would be distributed. Both of these problems inhibited major pharmaceutical companies from investing in vaccine research. In addition, there were and continue to be real questions about the possibility of producing a vaccine: scientists have been trying to develop an effective malaria vaccine for more than 50 years with little success.

The challenges to developing an effective vaccine are great. There are four species of malaria that infect people, and each species is composed of a number of genetically different strains. Human genetic differences can also affect the level of immunity in response to a vaccine. In addition, during the course of malaria infection, the human host confronts four distinct life-cycle stages: the invading sporozoite injected by the mosquito, separate replicating stages in the liver and the blood, and the mature sexual stages. Each of these life stages presents fresh antigens (targets) to the immune system. Moreover, it is unlikely that a vaccine that targets a single stage will effectively reduce malaria infections, because the parasite is at each stage able to reproduce in very high numbers. A vaccine that attacks the sporozoite stage, for example, would have to eliminate every sporozoite, for if a single sporozoite enters the liver it can lead to the production of thousands of merozoites, which enter the bloodstream and attack red blood cells. There has never been a vaccine developed against a complex multistage parasite.

To make matters more difficult, the malaria parasite is not a stationary target for the human immune system. For example, in *P. falciparum* infections, the multiple cycles of red cell invasion and parasite reproduction that occur after the parasite invades the bloodstream result in changes in proteins of the antigens expressed by the parasite on the surface of the invaded red blood cells. The immune system and, therefore, any vaccine need to be able to respond to these variable targets. As a review of vaccine progress noted in 1997, "Lying ahead is the fundamental question of whether any vaccine can overcome the parasites' aptitude for immune evasion, for this is a successful organism whose natural habitat is the bloodstream of the human host."[19]

Nonetheless, many scientists around the world think that a vaccine is possible. As we have seen, people living in holoendemic areas who are repeatedly exposed to malaria develop a functional immunity that controls the parasites in their blood and protects them from severe disease. In addition, studies in animals and small-scale human clinical trials have shown that immunization with attenuated (weakened) parasites can stimulate an immunity that protects against a subsequent challenge of fully virulent and viable parasites. New knowledge derived from the sequencing of the *P. falciparum* genome, as well as that of the *Anopheles gambiae* mosquito, has opened up fresh avenues for research. Most importantly, since the launching of RBM new funding initiatives have stepped up research on potential vaccine candidates. For example, a partnership between the Malaria Vaccine Initiative, funded by the Bill and Melinda Gates Foundation, and GlaxoSmithKline Biologicals, the world's largest manufacturer of vaccines, has made possible a scaling up of research into a vaccine, which Glaxo-SmithKline has been developing since 1984. The initiative, by sharing the financial risk of the vaccine development with Glaxo, enabled the vaccine giant to move forward at an accelerated pace.

It remains to be seen how long it will take to develop an effective vaccine. Most estimates put the horizon at around 2015. And then the likelihood is that we will have a vaccine that will not protect against malaria entirely but a targeted vaccine that will protect young children or help block transmission of the disease.

Reductions in Malaria?

Roll Back Malaria was successful in significantly reducing the burden of malaria in a diverse set of countries including Eritrea, Cambodia, and Vietnam. In most of sub-Saharan Africa, however, progress in limiting malaria was slow. As 2005 neared an end, it was clear that the majority of African countries were far from achieving the goals of the Abuja conference. According to the WHO's *World Malaria Report* for 2005, an average of just 3 percent of children younger than five slept under ITNs. The percentage of children with fevers being treated presumptively with antimalarials was better, with a regionwide average of 49.6 percent.[20] However, 95 percent of these treatments used chloroquine, for which there was widespread resistance. The coverage of pregnant women with intermittent preventive treatment (IPT) was difficult to measure. Surveys from Kenya, Ghana, and Zambia indicated that only 11 percent of women were receiving such treatment in 2003. A more recent study of the feasibility of incorporating IPT through routine antenatal care in Africa concluded that only half of the countries would reach the Abuja goal of 60 percent coverage.[21]

In reading these numbers we need to keep in mind two facts. First, acquiring accurate information on the distribution of RBM interventions was plagued by limited data. With few base-line surveys and only irregular data collection, it was difficult to determine to what extent RBM was making progress and whether any particular country had achieved the Abuja goals. Second, even in the limited data available, there was clearly a great deal of variation in the distribution of these interventions across countries as well as within countries. The sources of this variation are worth examining, for they reveal something about the problems inherent in the RBM strategy. This is particularly apparent in the data on the distribution of ITNs, arguably the central most important component in the RBM strategy in Africa.

In countries such as Tanzania, Kenya, and Nigeria, where ITN distribution initially followed the market model proposed by RBM partners that required individual families to purchase nets, ITN distribution remained low and uneven. For example, the 2005 Demographic and Health Study for Tanzania found that only 16 percent of young children and 16 percent of pregnant women slept under an ITN the night before the survey.[22]

Surveys revealed a positive correlation between socioeconomic status and ITN use, with the lowest levels of usage occurring among the poorest segments of the population. Not surprisingly, when queried as to why they did not purchase nets, the number one response given was the cost.[23] A five-dollar net may seem inexpensive. But, as we saw in chapter 7, in a country like Zambia, where the per capita annual income is less than a dollar a day, five dollars was a huge amount. While social marketing strategies had some success in creating demand for purchased nets, they did not overcome the problem that some groups were simply too poor to afford nets. Failure to recognize the role of poverty in limiting the success of health provision models that required individuals to pay for their own protection played a significant role in limiting the achievement of the Abuja goal of 60 percent ITN coverage.

By contrast, much higher levels of distribution were achieved where programs treated ITN distribution as a public health intervention on par with vaccinations. This meant distributing nets free or at highly subsidized rates, often as part of a wider health campaign involving the delivery of vaccinations and other health interventions. In Zambia, where fewer than 5 percent of children younger than five slept under ITNs in 2002, a project funded by the Canadian Red Cross, UNICEF, and the Canadian International Development Agency combined ITN distribution with measles vaccinations, vitamin A supplementation, and mebendazole treatment for intestinal worms in five underserved districts. According to a survey conducted after the campaign, this resulted in greater than 80 percent coverage for all interventions in the five districts, which had 89,000 children under five years of age. The delivery cost per ITN was only US$0.36. Kenya has recently increased net coverage dramatically by making nets available for purchase at highly subsidized prices. While the commercial price for nets remains between $4 and $6, they are available through health clinics for $0.60 apiece. According to Population Services

International, a nongovernmental organization that has played a major role in the social marketing of ITNs in Africa, the percentage of pregnant women and children under five sleeping under nets in Kenya increased from less than 20 percent in 2003 to nearly 50 percent in 2005 in three provinces.[24]

Similarly, in Eritrea, ITNs were distributed widely through the country's primary health care system. Initially, the program charged a small amount for each net. However, beginning in 2002, it began distributing nets for free in the most affected areas. They also instituted a free retreatment program centered at local clinics. Net usage rose dramatically. By the end of 2003, Eritrea had more than 60 percent of children under five sleeping under ITNs, making it the first country to achieve the Abuja goal.[25] Jeffrey Sachs deserves credit for pushing donors and RBM partners to abandon the self-financing aspects of ITN distribution and make them freely available to those at risk.

Yet, even where nets were made widely available, it remained unclear to what extent they were properly used and maintained. Studies done in Nigeria and Ghana indicated that the percentage of children under five sleeping under ITNs was significantly lower than the percentage of households possessing ITNs.[26] This suggested that adults used the nets for their own comfort or protection. Similarly, a follow-up study of ITN use in western Kenya, three years after nets were distributed to 60,000 households, found that most households had retained and repaired their original bed nets. However, none had acquired new nets or retreated their original nets with insecticide since the intervention was completed. Failure to retreat nets significantly reduces their protective value. Researchers in Kenya noted that the protective efficacy of ITNs in children younger than one year fell from 26 to 17 percent when retreatment was delayed beyond six months.[27] Finally, studies in the 1990s by anthropologist Peter Winch and his associates in Tanzania found that bed net use followed indigenous understandings that linked the risk of fever with rainy seasons. People in the studies tended to use the nets during this period but abandoned their use in subsequent months when the density of vectors and risk of infection were in fact greater.[28] There clearly was a need for more work on the demand side. Getting nets out there was not enough.

Success in insuring that young children received adequate malaria treatment within 24 hours of the onset of a fever was also uneven up through 2005. Standing in the way of achieving higher levels of coverage was the breakdown of government-funded health services in the 1980s and 1990s. As we saw in chapter 7, drug shortages resulting in temporary stock outages of front-line antimalarials were widespread in many African countries. Local government health services were also hampered by the absence of diagnostic equipment and personnel. Thus, even if a mother recognized that her child's fever required medical attention, the chance that the child would receive adequate treatment was limited by weaknesses in the health infrastructure.

An additional barrier to rapid treatment was the implementation of health care models that required patients to pay for services in order to reduce government expenditures during this same period. As with the practice of selling bed nets, the impositions of user fees for health services limited access for the poor.[29] As we saw in the case of Zambia, there was clear evidence that user fees discouraged poor mothers from seeking timely medical care. Efforts to develop mechanisms for exempting the poor from user fees for primary care did not work very well.

In a number of countries, designated first-line antimalarial drugs, most commonly chloroquine or SP, were distributed without charge. However, with widespread resistance to chloroquine and growing resistance to SP, children often required additional drugs for which clinics charged fees. The movement to more expensive combination therapies and artemisinin-based compounds made it increasingly difficult for government health services to provide free treatments.

Efforts to improve the early treatment of childhood fevers under the WHO's Integrated Management of Childhood Illness programs had some success in increasing the effective treatment of childhood malaria and preventing needless deaths. Yet, where health services remained understaffed and undersupplied, and user fees continued to be imposed, structural barriers inhibited the achievement of the Abuja goals. In Uganda, health authorities recognized the difficulty of reaching the Abuja goals for rapid treatment in an environment where access to effective medical treatments was limited by cost. The Ministry of Health, supported by the World Health Organization, was successful in reaching the Abuja target when it decided to start a program of free distribution of unit-dosed, prepacked antimalarial treatments (combination of chloroquine and SP) for children under five years of age through communities and the public health sector in 10 selected districts. The first follow-up survey in July 2003 indicated a significant improvement in timeliness and accessibility of adequate malaria treatment in the target population with close to 60 percent of children under five in the implementation areas receiving treatment within 24 hours.[30]

Preventing malaria in pregnancy may be easier than protecting children with ITNs or rapid treatment programs. Antenatal care programs have been highly successful in reaching women of all classes in both rural and urban settings throughout much of Africa. Regionwide program attendance averaged between 70 and 80 percent. Thus gaining access to women during pregnancy was not a problem. Yet neither the provision of intermittent treatment with antimalaria drugs nor the use of ITNs by pregnant women increased significantly. While ITN usage among pregnant women was handicapped by the same problems that limited their use to protect young children, low levels of intermittent preventive treatment in most African countries was caused by a combination of supply shortages, disagreements among international health authorities regarding the precise regimen that should be employed to protect women in pregnancy, lack of adequate training of local staff in the management of IPT, and local resistance to the preventive use of chloroquine or SP. It was frequently difficult to convince women who were asymptomatic that they should take preventative drugs. The use of chloroquine for prenatal protection was also difficult to understand in a region where the drug was widely used in high doses as an abortive agent. Many programs also found it a challenge to convince women to seek antenatal care early and often enough to effectively administer IPT. The country that had the greatest early success in implementing IPT programs was Malawi, which was a site of early testing of the effectiveness of IPT. The country adopted the program as part of its national malaria program even before the initiation of Roll Back Malaria and received substantial support for the program from outside nongovernmental organizations as well as from the Centers for Disease Control. Surveys indicated that 67.5 percent of pregnant women in 2004 received one preventive dose with SP. The percentage receiving the recommended second dose, however, was only 39.3 percent.[31]

As we have seen, most of the RBM interventions required behavioral change by populations at risk. In order for this to happen, RBM needed to involve whole communities in developing educational programs to inform people about what they needed to do to protect themselves and their children from malaria. This occurred very unevenly in most RBM-funded countries.

BOOSTING ROLL BACK MALARIA, 2004–2006

By the end of 2004, the World Bank together with its RBM partners recognized that the 2000 Abuja conference goals were unlikely to be achieved by 2005. The Bank's analysis of the problem raised a series of programmatic issues. Progress was limited by weak institutional capacities for procurement and financial management, inadequate health staffing, lack of diagnostic equipment, limited financial resources to support RBM interventions, and in many cases the absence of sufficient political leadership at the highest levels of government. The Bank's analysis pointed as well to the absence of adequate monitoring mechanisms for tracking and assessing program development and impacts. The Bank also recognized its own failings, both in directly funding RBM programs and in leveraging financial support from local governments and other external donors. In response to these problems, the Bank initiated a Booster Program for Malaria Control in 2005. Through this program the Bank committed itself to mobilizing "financial and technical resources from within and outside the institution, including the public and private sectors, to stimulate the production of commodities such as insecticide-treated bed nets (ITNs) and antimalarial drugs; lower taxes and tariffs on such commodities; improve and maintain long term commitment to malaria control by governments and civil society groups; and build public-private partnerships for program design management and evaluation."[32] The Bank and its RBM partners evaluated African countries in terms of their readiness to scale up malaria interventions. Several countries were identified as likely targets. Among the first of these was Zambia.

In 2005 the World Bank responded positively to a request from the Zambian government for US$20 million to be drawn from the US$350 million credit that Zambia had received from the International Development Association for the 2005–8 fiscal cycle. The funds were to be disbursed over three years beginning in January 2006 and were intended to help Zambia implement its 2006–11 National Malaria Strategic Plan for Malaria Control. The specific objectives of the plan were to increase the percentage of children who sleep under treated bed nets from 30 to 60 percent; to increase the percentage of women who receive a complete course of IPT from 45 to 70 percent; and to increase the percentage of people with fever who receive effective treatment within 24 hours of onset of illness from 60 to 80 percent. In addition, the booster program included the use of insecticides for spraying houses in districts where malaria is unstable. The goal was to increase the percentage of people in districts eligible for indoor residual spraying who sleep in appropriately sprayed structures from 40 to 80 percent. The overall goal of the booster program was to exceed Abuja goals and reduce malaria mortality by 75 percent by the end of 2008, or seven years ahead of the original RBM deadlines.[33]

The World Bank was not alone in its efforts to scale up malaria control in Zambia. It partnered with Gates Foundation–supported Malaria Control and Evaluation Partnership in Africa (MACEPA). In addition to supporting scaling up, MACEPA aimed to increase the administrative capacities of participating African countries, and to monitor and document program successes, including the economic impact of scaling up malaria control. In addition to the funds provided by the World Bank and the Gates Foundation, the Global Fund for AIDS, Tuberculosis and Malaria made grants of US$38 million and the government of Zambia committed US$7 million, some from government revenues, the rest made available through debt relief under the World Bank's Heavily Indebted

Poor Countries initiative. Finally, in 2006, the U.S. government through its Agency for International Development donated insecticides worth $827,000 to Zambia for malaria control.

Zambia was intended to be a model for the rest of Africa. If the scaling up of RBM worked in Zambia, the World Bank and its partners hoped that the Zambian success would serve as testament to what could be achieved through RBM with sufficient commitment and funding. On the other hand, if the National Malaria Control Programme failed to achieve its goals in Zambia, it would raise serious questions about the feasibility of the RBM strategy on the part of both international donors and national governments. As a public health specialist involved in the World Bank's malaria program commented. "If we are not successful, we may return to the dark ages of malaria funding."[34] A lot was at stake in Zambia.

The success of efforts to scale up malaria control will still be in the balance when this book is published. We are therefore left with a series of as-yet-unanswered questions. The most immediate questions are whether efforts to scale up malaria control in Zambia will succeed by 2008. Will malaria mortality be reduced by 75 percent? But an equally important question, given what we know about the history of past efforts to control malaria, is whether malaria control can be maintained in a resource-poor environment like Zambia over the long term. Can malaria control be maintained without a substantial improvement in the overall social and economic welfare of the Zambian population?

There are certainly reasons to be cautiously optimistic about the possibility of scaling up the distribution of RBM interventions. As noted earlier, a Canadian-funded project was able to achieve 80 percent coverage with heavily subsidized insecticide-treated bed nets in five districts of the country. It seems highly likely that similar successes can be achieved nationwide with the funding now available. A National Malaria Indicator Survey conducted under MACEPA guidance in 2006 found that 50 percent of households possessed ITNs, double the percentage reported three years earlier. It is less clear whether high levels of coverage can be translated into high levels of use, and in fact the survey found that only 25 percent of children were sleeping under ITNs.

Ensuring higher levels of coverage for children will require additional investments in health education, community participation, and monitoring.

Similarly, antenatal care coverage is high in Zambia, with an estimated 90 percent of expectant mothers seeking professional health care in 2001–2.[35] Thus efforts to increase intermittent preventive treatment to protect women from malaria during pregnancy should be possible. However, the experience of a number of African countries with similar levels of antenatal care usage indicates that the problem is not getting women to access care but getting them to access it early and often enough during their pregnancy to implement intermittent preventive treatment effectively. So achieving high levels of coverage will require efforts to educate women as well as to provide increased access to drugs.

Ensuring the rapid diagnosis and treatment of children is problematic largely because of weaknesses in the country's health system. Fifty percent of Zambia's health facilities were closed in 2005 due to staffing shortages, and only half of the country's rural population resided within three miles of a health clinic. In many rural areas the distances were much greater. The limitations of the current health infrastructure were laid bare by the government's decision to eliminate health care user fees in rural areas on April 1, 2006. The decision followed the summit meeting of the G8 countries in July 2005, at which member countries agreed to reduce Zambia's external debt by US$4 billion. The decision to eliminate user fees resulted in a sharp increase in clinic and hospital attendance. Without any increase in government health personnel or facilities, however, the flood of patients overwhelmed the health system and resulted in long lines and drug outages. Thus, aside from making sure that effective antimalarial drugs are available at an affordable cost, that parents are able to recognize the clinical signs of malaria, and that they are willing to seek rapid medical treatment for their children, the partners involved in scaling up malaria control will need to help Zambia invest in broad-based improvements in the country's health infrastructure. This will involve recruiting and training new health care providers to replace those who have dropped out of the system when their salaries were cut, and raising salaries to levels that will ensure their retention.

Ironically, it was the World Bank's conditionalities during the 1990s that forced the government to cut health workers' salaries and created the health care shortages the country currently faces. Many of the rural health facilities will also need to be repaired, and the overall network expanded. Evaluating health sector development in Zambia, in August 2005, Takatoshi Kato, deputy managing director of the International Monetary Fund concluded:

> The health system faces severe constraints that prevent many public health needs from being adequately addressed. Staffing levels are insufficient—a problem exacerbated by the massive emigration of health workers to wealthier countries. Donors' response to the capacity problem in health care has been, in part, to set up parallel delivery systems that attract workers from the public sector, thereby bypassing the urgent need for capacity building in that sector. However, to raise staffing levels and equip health centers, the government and its development partners must make major investments.[36]

It is not clear to what extend the booster program partners, in their rush to meet self-imposed deadlines for the distribution of malaria interventions, are also prepared to support health system strengthening. As of November 2006, a great deal more needed to be done.[37]

Rapid diagnosis and treatment may be facilitated by the wide distribution of rapid diagnostic tests and combination therapies that reduce treatment times. In addition, the use of commercial outlets to make drugs available could contribute to scaling up (though commercial outlets would be effective in reaching the poorest populations only if the cost of drugs were subsidized and the distribution networks were monitored to ensure quality). There is also the inherent danger that reliance on commercial networks to deliver drugs and nets would deflect governmental efforts to strengthen the country's health system.

Although the prospects for scaling up RBM in Zambia seem generally good, the likelihood that high levels of coverage can be sustained seems less certain. But some knowledgeable people are optimistic. Dr. Kent Campbell directs the MACEPA initiative for the nongovernmental organization PATH. He is a man who knows a great deal about malaria control. Before joining PATH he was director of the Vector Disease Division of the Centers for Disease Control and has been involved in combating malaria in many parts of the globe for more than 25 years. He is convinced that the disease can be controlled in Zambia and that, with the support of the World Bank, the Global Fund for AIDS, Tuberculosis and Malaria, and the Gates Foundation, Zambia will be a model for the rest of Africa. I once asked him whether malaria control could be sustained in Zambia after the scaling-up project was completed. His initial reply was, "There are two words that I cannot stand. The first is colonialism, the second is sustainability. The first is used as an excuse for the problems developing countries face. The second is used as an excuse to do nothing."

There is no doubt some truth in this statement. While Zambia's social and economic problems had their roots in patterns of uneven development and reliance on copper during the colonial period, much went wrong after independence. Arguments that we need to be able to sustain programs before we begin them, while not necessarily an excuse for inaction, can be used to delay and limit action. Nonetheless, there is no way to get around the fact that the Zambian malaria control project has to overcome the effects of political and economic maldevelopment that began before independence. It is also a reality, as Campbell himself acknowledged, that sustainability is a very real concern. As we saw in chapter 7, reductions in malaria morbidity and mortality achieved by the end of the Malaria Eradication Programme were lost in many parts of the world during the 1970s and 1980s when malaria programs failed to sustain control. They were also lost because malaria eradication had done little to alter the ecological, social, and economic forces that drove malaria transmission in many parts of the world.

More recently, the World Bank's proclaimed success in greatly reducing malaria in the Amazon region of Brazil through its Amazon Basin Malaria Control Project was reversed. The US$73 million project reduced reported cases by 60 percent from 557,787 to 221,600 between 1989 and 1996. The decline was not sustained, however, and from 1998 to 1999 there was even a 34 percent increase in the number of malaria cases.[38] The issue of sustainability is also raised by one of RBM's much vaunted success stories: Eritrea. This country, with substantial support from RBM partners, including the World Bank, WHO, the U.S. Agency for International Development, and the Norwegian Agency for Development Cooperation, nearly halved the proportional malaria mortality rate among children under five, slashing the number of fatalities to one of the lowest levels in Africa between 1999 and 2004. Overall malaria morbidity was reduced by 80 percent.[39] However, ongoing border conflicts between Eritrea and Ethiopia along with alleged human rights abuses have led outside donors to largely abandon Eritrea for political reasons. Of the various agencies that have supported malaria control efforts, only the World Bank remained at the end of 2006, with an investment of $400,000 per year.[40] In its Project Appraisal Document in 2005, Bank officials concluded that Eritrea's interventions in "HIV/AIDS/STI, TB, Malaria and Reproductive Health" could not be financially sustained without continued external support. The country therefore faces the very real possibility that the improvements in malaria control achieved over the past five years will be lost.[41]

Failure to sustain high levels of malaria protection could have serious health consequences for populations left unprotected. This is particularly true for populations living in holoendemic regions of Africa, where repeated exposure to malaria produces an acquired functional immunity to the disease. This immunity is not permanent and can be undermined or lost if exposure is interrupted, as is likely to occur if transmission is significantly reduced for several years. In such circumstances, the protected population would be highly vulnerable to infection if control measures broke down, as the disastrous consequences in Madagascar, described in chapter 6, indicate.

The question of whether populations with acquired immunity to malaria should be included in malaria control programs has been a contentious subject for almost a century.[42] Most recently the question has arisen in relation to the use of insecticide-treated bed nets in Africa. A study conducted in Tanzania on the impact of ITN use among children under four concluded that the immune responses to an important target of protective malaria immunity,

VSA antibodies, were influenced by bed net usage and suggested that ITNs in the longer run could influence malaria disease patterns. This study suggested that children protected by nets would have a reduced level of immunity in later life. They would also be at greater risk from malaria should protection levels decline.[43] Other studies comparing clinical data, however, found no evidence that bed net use increased morbidity in later years.[44] Kent Campbell rejected arguments about lost immunity. Yet he acknowledged that we currently do not know how immunity will be affected with the levels of protection that RBM will provide. He also noted that immunity levels will be monitored in Zambia.

There are reasons to be hopeful that a malaria control program in Zambia can be sustained. To begin with, the costs of maintaining control are likely to be less than the costs of scaling up. The initial distribution of nets will cost considerably more than their retreatment. Moreover, if 80 percent of the population is protected with nets or spraying, the epidemiology of the disease will change. Transmission will be reduced. Fewer people will need treatment. That means the high cost of providing effective treatment with artemisinin-based combination therapies will be reduced. It is also possible that IPT programs will no longer be necessary. Second, a successful program may convince Zambian leaders, as well as outside donors, of the value of maintaining a long-term commitment to control, especially if it can be demonstrated that controlling malaria has economic benefits, as RBM and the booster advocates maintain.[45]

On the other hand, there are also reasons to question whether control can be maintained. It is unclear if treated nets will remain effective against anopheline mosquitoes or whether mosquitoes will develop resistance to the pyrethroid insecticides used to treat them. Researchers found as much as 80 to 90 percent pyrethroid resistance in *Anopheles gambiae,* the primary malaria vector in some parts of Côte d'Ivoire, Benin, and Burkina Faso, due to mutations in the gene that encodes the pyrethroid target protein in the mosquito's nervous system.[46] On the other hand, studies in areas with a very high proportion of pyrethroid-resistant *A. gambiae* mosquitoes showed that pyrethroid treated bed nets still reduced malaria inoculation rates and the incidence of malaria episodes in children. The researchers warned that other forms of resistance might develop, however, reducing the effectiveness of ITNs, and urged both continued surveillance and further research.[47] Some investigators have developed nets with multiple toxins to retard the development of resistance in the same way that combination therapies reduce drug resistance.[48] The need to develop and distribute more effective nets in the future could increase the costs of maintaining control over time. So, too, would the shift to long-acting insecticide treated nets designed to overcome the problem of retreatment. The long-acting nets cost about double the cost of current ITNs. However, the cost over the life of the two nets may be comparable.

Similarly, recent reports (as yet unconfirmed) of *P. falciparum* resistance to artemisinin in West Africa remind us that providing rapid effective treatment for malaria, as well as protecting pregnant women with IPT, may require the continued development of new drugs. The possibility of artemisinin resistance was heightened by the willingness of pharmaceutical companies to market the drug as a stand-alone therapy, rather than in combination with other antimalarial drugs, as the WHO recommended.[49] The increasing costs involved with each new generation of antimalarial drugs threaten to further increase the price of control and the ability of countries to sustain control programs. The development and adoption of artemisinin-based combination therapies has already greatly increased the cost of drug treatment.

More broadly, the long-term maintenance of malaria control requires political support at the highest levels of government. While the current president of Zambia strongly supports malaria control, it is uncertain whether this position will be shared by his successor. One of the lessons of malaria control over the past 50 years has been that as malaria recedes, it increasingly becomes a disease of the poorest, least visible portion of the population. When this happens, malaria control becomes a lower priority for those who control the reins of government. Control activities are neglected, allowing for the reemergence of pockets of transmission that can quickly spark larger regional epidemics. This is especially true where populations move back and forth between poorer and better-off regions of a country in search of employment, as is the case in Zambia.

The heavy involvement of MACEPA consultants in overseeing the present scaling-up process also raises questions about sustainability. Can control be maintained without this external expertise? Have sufficient energy and resources gone into ensuring that the personnel and infrastructure needed to maintain control will be in place when the external advisers leave? Will the drive to meet a timetable force MACEPA to outsource administrative functions, such as procurement, at the expense of capacity building?

Finally, sustainability will be threatened unless efforts to control malaria are paralleled by a broader commitment to improving the economic and social well-being of Zambia's population. While the Zambian economy improved from 2003 to 2006, thanks to a combination of additional debt relief and a rise on copper prices, the country remained impoverished by international standards and continued to exhibit many of the economic and social problems described in chapter 7. Zambia ranks 90 out of 103 developing countries on the UNDP's Human Poverty Index.

MACEPA director Kent Campbell acknowledged that Zambia currently lacks the economic resources and administrative infrastructure needed to sustain the high levels of coverage that will be achieved through scaling up. Nor can the majority of the population, who earn less than one dollar per day, afford to maintain their own protection over the long term. In the near term, sustained control will require the continued commitment of foreign donors.

In the long term, however, the history of international support for malaria control strongly suggests that Zambia, and other countries fighting malaria, cannot depend on donor countries and organizations to maintain high levels of support. Donor fatigue and pressure to fund other programs eventually erode support for malaria control. Long-term sustainability, therefore, will require higher levels of economic growth, the elimination of poverty, and an overall improvement in the economic status of Zambia's population. These developments will allow the government to rebuild its health system and fund basic social services. They will also provide Zambians with the opportunity to earn the resources they need to protect themselves from infection through properly constructed houses, adequate diets, bed nets, and medicines. In the end, countries like Zambia need to "grow out of malaria." This does not mean that economic growth will eliminate malaria, especially in Africa. But it will make it possible for governments and individuals to take and sustain actions that will reduce the burden of the disease.

Reducing the burden of malaria may by itself help Zambia achieve these goals. It will reduce the medical expenses borne by individual families, along with the heavy burden that malaria imposes on the country's health budget. Individual enterprises may profit from a healthier work force. Children may suffer fewer absences from school. Individual workers may suffer fewer sick days. All of this is for the good. But the alleviation of poverty will require more long-term efforts to diversify the economy, reduce the country's reliance on copper, expand agricultural production, encourage the development of secondary industries, and continue to reduce or eliminate the country's debt burden.[50]

None of this will be easy. Moreover, it will require more than throwing money at Zambia. It will require, among other measures, reducing agricultural subsidies in Europe and America. As we saw in the case of cotton, these subsidies have suppressed agricultural prices, restricted efforts to diversify the economy, and limited the income of the rural poor. If more rural families can earn income from farming or other forms of employment, they will be better able to purchase nets and medications. In the end, efforts to scale up malaria control need to be viewed as a measure that will save lives in the short and middle term until malaria-affected countries can achieve a level of social and economic development that will sustain the general health of their populations.

The question, then, is whether national governments and external donors have the will and the ability to sustain the recurring and potentially increasing costs of malaria control until infected countries are able to grow out of malaria. To what extent will the competing development demands redirect malaria funding? Will the international donor community be willing to sustain the support, not only for malaria control, but also for economic policies that will ensure real long-term economic growth and development? Will the U.S. government be willing to limit its support for Texas cotton growers in order to aid farmers in Zambia? The answer to these questions will help determine the future of malaria and whether it will remain a tropical disease.

Ecology and Policy

A central argument of this book has been that the history of malaria has been driven by the interplay of social, biological, economic, and environmental forces. The shifting alignment of these forces has largely determined the social and geographical distribution of the disease, including its initial global expansion, its subsequent retreat to the tropics, and its current resurgence. By contrast, efforts to control malaria since the end of the nineteenth century have been largely driven by a narrower vision of the disease and its causes that has privileged biological processes and focused on attacking anopheline mosquitoes and malaria parasites. This biological model has informed public health policy since the discoveries of Lavaran, Ross, and Grassi and is at the heart of current Roll Back Malaria strategies. Put simply, malaria policy has largely ignored the human ecology of malaria. The failure to link ecology and policy has prevented the elimination of malaria as a serious public health problem in many areas of the globe and will continue to restrict the success of malaria control programs, such as Roll Back Malaria.

The tension between ecological approaches to disease prevention and control and narrower biomedical strategies is by no means limited to the problem of malaria. It is a fundamental dilemma of modern public health. Does one develop targeted strategies for attacking the immediate causes of disease, or focus on the broader societal determinants of ill health? Arguments for the former have rested largely on issues of efficacy, cost effectiveness, and what might be called "doability": we have the tools to greatly reduce suffering from particular diseases by eliminating the sources of infection and should employ them. Ecological arguments have countered that targeted approaches, while saving lives in the short term, fail to address the underlying forces driving disease and disability. Control therefore needs to be sustained indefinitely. Only by improving the overall well-being of a population can good health be achieved and sustained. As suggested in chapter 5, this debate has been a recurring feature of international health policy discussions since the end of the nineteenth century.

Since the 1980s the pendulum has swung decidedly toward narrower biomedical approaches aimed at eliminating the immediate causes of disease. This shift has been driven by a combination of cost constraints on health programs during the 1980s, the dominance of neoliberal thinking about health reform and the World Bank's role in international health funding since the early 1990s, the global health agenda of the Bill and Melinda Gates Foundation, and new efforts to mobilize support for disease eradication campaigns. The recession of the 1980s made broader approaches to health, as envisioned by advocates of the primary health care movement in the 1970s, too costly to pursue. Neoliberal policies that favored private-sector participation in health and the commodification of health interventions, together with the World Bank's econometric approach to health, have encouraged the creation of narrowly defined cost-effective strategies that produce measurable results. The Gates Foundation has poured hundreds of millions of dollars into the development of new medicines, vaccines, and targeted strategies. Supporters of new eradication campaigns against polio and measles, who have invoked the success of the smallpox eradication campaign of the 1960s and 1970s, have further encouraged the belief that diseases can be conquered by technology, without addressing the broader social determinants of health.

By emphasizing the important role played by ecological forces in the history of malaria, this book has attempted to push the pendulum back toward the center. I am not arguing that malaria control strategies are ineffective, or that we need to redirect all our efforts to dealing with the ecological forces driving malaria. Roll Back Malaria interventions, if scaled up, can significantly reduce the burden of malaria. We therefore need to insure that populations at risk, particularly the poorest and most vulnerable among them, can obtain the drugs, nets, and, hopefully one day, vaccines that will protect them from the ravages of malaria. We must also ensure that the acquisition of these resources can be sustained. In the short run, so much of the problem, as Paul Farmer so eloquently puts it, is "access, stupid."

In the long run, the social and economic conditions that drive malaria transmission—including patterns of labor exploitation that place workers at risk of infection, warfare that disrupts local ecologies and health services, population displacements that have exposed millions to malaria, and poverty that prevents individuals and countries from achieving and sustaining adequate levels of malaria protection—need to be reduced or eliminated. Real economic development that leads to broad-based improvements in the quality of life, as opposed to growing disparities between haves and have-nots, needs to be actively encouraged. Countries, particularly in Africa, need to be given the chance to grow out of malaria.

If we do not address the underlying ecological forces driving malaria, we will be condemned to providing protections against the disease indefinitely. For, as we have seen, any slackening of control will allow the forces driving malaria to rekindle the disease. The history of malaria over the past 50 years is strewn with examples of control programs that have broken down, allowing a resurgence of the disease. The ability of public health programs or individuals to maintain effective malaria control, moreover, is likely to become increasingly difficult and expensive given the propensity of both malaria parasites and anopheline mosquitoes to develop resistance to drugs and pesticides. In the long run, addressing the ecological determinants of malaria will reduce and in some cases eliminate the need to continually protect populations at risk. In those areas where malaria is unlikely to disappear, including much of sub-Saharan Africa, broader-based improvements in social and economic welfare will allow those at risk to protect themselves.

Those charged with developing strategies for combating malaria, therefore, need to take seriously what is currently only rhetoric in Roll Back Malaria. Malaria is a multisectoral problem that needs to be attacked on multiple fronts. Here, it is worth quoting again from the 1992 Global Malaria Strategy, upon which the Roll Back Malaria Partnership was based:

> Malaria control is not the isolated concern of the health worker. It requires partnership of community members and the involvement of those involved in education, the environment in general, and water supply, sanitation and community development in particular. Malaria control must be an integral part of national health development and health concerns must be an integral part of national development programmes.[1]

It is imperative that malaria control be part of a broader national development strategy. Moreover, as we saw in Zambia, the national economic problems of many developing countries in Africa cannot be solved from within. Internal adjustments to ensure economic growth must be complemented by increased access to international commodity markets, the reduction of unfair agricultural subsidies in developed countries, the opening up of European and American markets to African manufactured goods, and continued reductions in debt payments that eat up national development budgets. The new globalized economy needs to be a two-way street, not just a means of opening up African and other tropical markets to the products of northern industrial manufacturers and farmers.

Developed nations also need to take a more active role in both preventing and limiting armed conflicts that disrupt economies, destroy health services, and contribute to the loss and displacement of millions of lives. The human tragedies of civil wars in Darfur, Rwanda, Cambodia, Tajikistan, and Colombia—to name but a few—have all been made worse by the unleashing of malaria epidemics.

This is not an either-or choice. The interventions of Roll Back Malaria, together with research into new technologies for preventing and curing malaria, need to be developed in partnership with efforts to address the ecological conditions that produce malaria.

Without this kind of broad-scale, multisectoral approach, without a commitment to integrating ecology and policy, malaria will remain a threat to the health of millions of people living across the globe.

In the end, I return to Mulanda and I think of the parents who came to the clinic too late to save their feverish children from malaria, or who came but found that the medicines they needed were not available. I also remember the woman with trachoma who interrupted her treatment to attend her mother's funeral in Busoga, only to be reinfected with the disease while there. It is easy to think that had the parents acted differently, their children would have survived. If the woman had continued her treatment she would have been cured. But the realities are not so simple. The decisions these people made were shaped by the contours of the world in which they lived. Given access to an efficiently run health system and the resources to access it, they could have chosen the path to better health. Without those resources, that path was too often blocked. With adequate income they might have avoided their infections altogether. Without those resources, disease frequented their lives. We must not fool ourselves into believing that we can reduce the "intolerable burden of malaria" by simply finding better ways to attack anopheline mosquitoes and malaria parasites. We must also find ways to remove the barriers that prevent millions of people from achieving and maintaining health. If we do not, malaria will remain a recurring challenge to the health of the people who live in places like Mulanda.

ACKNOWLEDGMENTS

Writing a book about the history of malaria is not easy. It can take years to understand the life cycles of the four human malaria plasmodium, the intricacies of the breeding and feeding habits of the various species of anopheline mosquitoes capable of transmitting malaria, and the impact of environmental and climatic change on the reproduction of malaria parasites and the capacity of anopheline mosquitoes to transmit the disease. Added to this is the need to understand the broader social and economic contexts within which malaria occurs. People have spent their entire professional careers trying to understand just one small piece of this amazingly complex disease. I did not have that much time. I was, however, fortunate to receive the help of many people who have spent their lives researching and combating malaria.

When I first began working on malaria in the early 1980s, Professor Fred Dunn, a malaria epidemiologist from the University of California–San Francisco, provided important guidance. I also benefited from the expertise of malariologists Andy Speilman and Andy Arata.

Later, when I began teaching at Emory University in Atlanta, I was fortunate to have the opportunity to work with a number of malaria specialists at the Centers for Disease Control and Prevention. I cotaught course modules on malaria in the Department of International Health with Trent Ruebush, Jesse Hobbes, Monica Parise, Frank Richards, Rick Steketee, and Carlos Kent Campbell. Much of what I know about malaria I learned from my interactions with these dedicated malaria professionals. I am particularly indebted to Kent Campbell, who appears in the last chapter of this book, and who kindly read portions of this manuscript. I also wish to thank anthropologist Peter Brown, who has worked extensively on the history of malaria in Sardinia, for sharing his knowledge of malaria and his friendship over my 10 years at Emory and beyond.

At Johns Hopkins I have benefited from the proximity of the Malaria Research Institute and its many researchers at the Bloomberg School of Public Health. I am particularly grateful to Clive Shiff, who has spent his professional career combating malaria in Africa. He kindly read an earlier draft of this manuscript and gently revealed its many shortcomings. Protik Basu, a graduate of the Bloomberg School who now works on malaria at the World Bank, generously provided insights into the workings of the World Bank and its Malaria Booster Program.

The history of malaria has become a kind of growth industry in the last 15 years, and I have benefited from meeting and sharing my work with a number of historians and others working in this field. These interactions were greatly facilitated by historians William Bynum and Bernadino Fantini, who along with malariologist Mario Coluzzi, created a History of Malaria Network and organized three conferences around this topic. The conferences were attended by both historians working on malaria and by malariologists and others who had been part of this history. The conferences resulted in three special issues of the Italian parasitology journal *Parassitologia*. I learned a great deal from my interactions with all of these scholars and practitioners. But I have benefited enormously from my continued interaction with several of them. Brazilian historian Paulo Gadhela, whom I met at the first conference in Bellagio, Italy, subsequently coauthored an article with me on the history of the *Anopheles gambiae*'s invasion of northeast Brazil and Fred Soper's campaign to eradicate the invader. Jose Najera, who for years directed the malaria program at the World Health Organization, and whom I first met while doing research at the WHO headquarters in Geneva in 1992, generously shared with me his immense knowledge of malaria and its history. I also want to acknowledge Socrates Litsios, whom I also met at WHO in 1992, and who over the years has shared his broad experience working on malaria and his encyclopedic knowledge of the field of international health. James Webb, who is currently working on a history of malaria, read and commented on chapter 1 of this book.

The knowledge I gained from my interactions with all of these people led me to believe that I could write a global history of malaria. Writing this history for an audience of nonspecialists has, however, required me to condense much of what I have learned and to simplify, or simply leave out, many aspects of malaria and its history. I apologize in advance to those who have given so generously of their time and knowledge for any errors or distortions that these editorial decisions have produced.

While I was writing this book, I benefited from advice of my colleagues in the Institute of the History of Medicine, who kindly read a draft of the manuscript and provided much needed feedback. Michael Henderson, a graduate student from the History Department, helped me sort through the masses of secondary literature on the social and economic history of the regions I discuss in the first three chapters of the book. Charles Rosenberg, who convinced me to write this book, and Jackie Wehmueller, my editor at the Press, read two drafts of the book and provided guidance and advice on how to write a book that would be both authoritative and accessible. I hope I have achieved this goal.

I also owe a special thanks to Susan Abrams, with whom I have had the privilege to work on the *Bulletin of the History of Medicine* since I arrived at Johns Hopkins. Susan generously agreed to give the manuscript one last read before it went to press. She applied her skilled editorial eyes to the manuscript and at almost every point made suggestions that improved its readability.

I have been fortunate over the years to receive generous research support for my malaria work from a number of organizations. My early work on malaria in Africa was supported by grants from the American Council of Learned Societies, the Fulbright-Hays Fellowship Program of the Council for the International Exchange of Scholars, and the Department of Education. Grants from the National Endowment for the Humanities, the National Library of Medicine, and the Rockefeller Archives Center supported my subsequent work on the history of international malaria control programs. I want to thank each of these organizations for the support.

❖ ❖ ❖

Finally, I want to thank my wife, Carolyn, to whom I have been married for 30 years and who has supported me in so many ways over the long years that I have worked on malaria. To her oft-repeated question, "Are you still working on that malaria book?" I can now answer, "No dear."

NOTES

Preface. Mulanda

1. Robert Snow et al., "Pediatric Mortality in Africa: *Plasmodium falciparum* Malaria as a Cause or Risk?" *American Journal of Tropical Medicine* 71 (2004): 16–24.

Introduction. Constructing a Global Narrative

1. Henry Sigerist, *Medicine and Health in the Soviet Union* (New York: Citadel Press, 1947): 165–66.

2. Lewis Hackett, *Malaria in Europe: An Ecological Study* (Oxford: Oxford University Press, 1937), 223.

3. *Pravda* reported that in 1921–22 famine touched 25 million people and led to the death of unknown millions.

4. W. Bruce Lincoln, *Red Victory* (New York: Simon and Schuster, 1989), 462–72.

5. Sigerist, *Medicine and Health in the Soviet Union,* 165–66.

6. The League of Nations Malaria Commission's "Report of Its Tour of Investigation in Certain European Countries in 1924," described the events preceding the epidemic in these terms: "The extraordinary spread of malaria in the whole of Russia, especially in the Volga district, was largely influenced by the European War, as well as by the Russian Civil War and all its consequences. The movement of armies and refugees, the destruction of the technical sanitary organizations, and the famine of 1921–22 contributed not only to the increase of malaria morbidity but also to the serious effect of the disease on the weakened population. Hunger forced the population to emigrate to Turkestan, Bukhara, Siberia and the Caucasus, whence the most severe forms of malaria were imported" (141).

7. Ira Klein, "Development and Death: Reinterpreting Malaria, Economics and Ecology in British India," *Indian Economic and Social History Review* 38, 2 (2001): 162.

8. Charles A. Bentley, *Malaria and Agriculture in Bengal* (Calcutta, 1925).

9. Changes imposed on India's water flows, salinity, and temperature quickly encouraged the proliferation of *Anopheles philippinensis, Anopheles sundaicus, Anopheles maculatus, Anopheles stephensi, and Anopheles culicifacies.* Klein, "Development and Death," 160.

10. Centers for Disease Control, "Local Transmission of *Plasmodium vivax* Malaria—Palm Beach County, Florida, 2003," *Morbidity and Mortality Weekly Review* 52, 38 (September 26, 2003): 909–11.

11. Richard Carter and Kamini M. Mendis, "Evolutionary and Historical Aspects of the Burden of Malaria," *Clinical Microbiology Reviews* (October 2002): 582.

12. Hackett, *Malaria in Europe*; L. J. Bruce-Chwatt and J. de Zulueta, *The Rise and Fall of Malaria in Europe* (Oxford: Oxford University Press, 1980); Angelo Celli, *The History of Malaria in the Roman Campagna from Ancient Times* (New York: AMS Press, 1977); Mary Dobson, *Contours of Death and Disease in Early Modern England* (Cambridge: Cambridge University Press, 1997); St. Julien Ravenel Childs, *Malaria and Colonization in the Low Country, 1526–1696* (Baltimore: Johns Hopkins Press, 1940); Erwin H. Acherknecht, *Malaria in the Upper Mississippi Valley, 1760–1900* (Baltimore: Johns Hopkins Press, 1945); Margaret Humphreys, *Malaria, Poverty, Race and Public Health in the United States* (Baltimore: Johns Hopkins University Press, 2001); Robert Sallares, *Malaria and Rome: A History of Malaria in Ancient Italy* (Oxford: Oxford University Press, 2002); Frank Snowden, *The Conquest of Malaria: Italy, 1900–1962* (New Haven: Yale University Press, 2006); Sheldon Watts, "British Development Policies and Malaria, 1897–1929," *Past and Present* 165 (1999): 141–81; R. M. Packard, "Maize, Cattle, and Mosquitoes: The Political Economy of Malaria Epidemics in Colonial Swaziland," *Journal of African History* 25, 2 (1984): 189–212; Ira Klein, "Development and Death: Reinterpreting Malaria, Economics and Ecology in British India," *Indian Economic and Social History Review* 38, 2 (2001): 162.

13. Benjamin Rush, "An Enquiry into the Cause of the Increase of Bilious and Intermitting Fevers in Pennsylvania, with Hints for Preventing Them," in *Transactions of the American Philosophical Society, Held at Philadelphia, for Promoting Useful Knowledge* (1786), 2:207.

14. Ironically, the patterns of industrialization and agricultural growth in Europe and North America were supported by the importation of raw materials from the tropical South, extracted under conditions of production that encouraged the transmission of malaria.

15. M. A. Barber, "The History of Malaria in the United States," *Public Health Reports* 44, 43 (1929): 2579.

Chapter 1. Beginnings

1. This group of parasites is part of a larger collection of parasites belonging to the phylum Apicomplexa that share many anatomical and life-cycle characteristics. The group may have had its origins as free-living single-cell organisms that became parasites on insects, and as some of these hosts adapted to feed on blood, the parasites transferred to the blood as a convenient source of nourishment.

2. For the case for a Southeast Asian origin, see G. R. Coatney et al., *The Primate Malarias* (Bethesda: U.S. Department of Health, 1971). More recently, see A. Ananias, A. Escalante, et al., "A Monkey's Tale: The Origin of *Plasmodium vivax* as a Human Malaria Parasite," *Proceedings of the National Academy of Medicine* 102 (2005): 1980–85. The case for an African origin can be found in Richard Carter and Kamini Mendis, "Evolutionary and Historical Aspects of the Burden of Malaria," *Clinical Microbiology Reviews* 15, 4 (October 2002): 575–76, and Richard Carter,

"Speculations on the Origins of *Plasmodium vivax*," *Trends in Parasitology* 19, 5 (May 2003): 214–19.

3. Mario Coluzzi, "The Clay Feet of the Malaria Giant and Its African Roots: Hypotheses and Inferences about Origin, Spread and Control of *Plasmodium falciparum*," *Parassitologia* 41 (1999): 277–83.

4. Robert Sallares, *Malaria and Rome: A History of Malaria in Ancient Italy* (Oxford: Oxford University Press, 2002), 24–35, has recently challenged this argument in an effort to show that *P. falciparum* is an ancient parasite in man. He states that genetic evidence of the early divergence of *P. falciparum* from its primate cousin *P. richenowi* is proof that *P. falciparum* could exist in small bands of hunters and gatherers. Yet he does not explain how this could have occurred given the high pathogenicity of the parasite. He sees the high pathogenicity of *P. falciparum* as an adaptation to an environment where host-to-host transmission was available all year long. In such environments "the type of parasite which will achieve the greatest evolutionary fitness is the one which achieves the highest rate of reproduction in the host, irrespective of what that does to the host." *P. falciparum* achieves this in a number of ways, including invading nearly all of the red bloods cells in its human host. By contrast, he argues, the less aggressive behavior of *P. vivax* and *P. malariae* was an adaptation to the human host's need to survive over cold winter months in order to have the opportunity for transmission to new hosts the following year. This may be partly true. However, there are several problems with this reasoning. First, the conditions under which transmission was possible year round depend on an adequate population of human hosts and efficient mosquito vectors. It is unclear that either of these conditions existed before the development of agriculture in Africa. Second, while *P. vivax* and *P. malariae* are well adapted to survival in cooler climates, we cannot explain their behavior as an adaptation to this condition, because the same behavior exists in Africa. Granted, there is evidence that with *P. vivax* speciation has led to increased frequencies of liver-stage parasites in cooler climates. Still, given the existence of similar levels of pathogenicity in Africa, one cannot credit their low level of pathogenicity to a selective adaptation to cooler climates unless one assumes an evolutionary history in which the two parasites originated in these cooler regions and then spread into Africa. This runs counter to both logic and genetic evidence. See Coluzzi, "The Clay Feet of the Malaria Giant and Its African Roots," 278; Jennifer C. C. Hume et al., "Malaria in Antiquity: A Genetics Perspective," *World Archeology* 25, 2 (2003): 180–92; and Stephan M. Rich, "Malaria Eve: Evidence of a Recent Population Bottleneck throughout the World Populations of *P. falciparum*," *Proceedings of the National Academy of Science* 95 (April 1998): 4425–30, for discussions of recent findings regarding genetic evidence of the evolution of *P. falciparum*. Evidence for the more recent development of *P. falciparum* is also supported by haplotype analyses of various genetic polymorphisms, such as G6PD and thalassemias, that provide protection from malarial infections. These analyses indicate that both mutations are of relatively recent origin, that is, less than 10,000 years old.

5. Lewis Hackett, *Malaria in Europe* (Oxford: Oxford University Press, 1937), 66.

6. Recent research suggests that gametocyte production may be stimulated by mosquito bites, emerging at the beginning of breeding seasons.

7. Hackett, *Malaria in Europe,* 66–67.

8. In the case of *P. vivax,* for example, the reproductive process takes 9 days at 25 degrees centigrade, but 30 days at 16 degrees centigrade. Ibid., 67–69.

9. J. D. Charlwood et al., "Survival and Infection Probabilities of Anthropophagic Anophelines from an Area of High Prevalence of *Plasmodium falciparum* in Humans," *Bulletin of Entomological Research* 87 (1997): 445–53.

10. Carter and Mendis, "Evolutionary and Historical Aspects," 577.

11. The intensity of malaria parasite transmission is normally expressed as the entomologic inoculation rate (EIR), the product of the vector biting rate times the proportion of mosquitoes infected with sporozoite-stage malaria parasites. Malaria transmission intensity in Africa is highly variable with annual EIRs ranging from <1 to >1,000 infective bites per person per year.

12. See Phillip Curtin, Jan Vansina, and Steven Feierman, *African History* (Boston: Little Brown, 1978), 7–11, and John Iliffe, *Africans, the History of a Continent* (Cambridge: Cambridge University Press, 1995), 12–17, for discussions of the origins of agriculture in Africa.

13. In this regard it is worth noting that *A. gambiae* is now recognized as a member of a complex that includes six species. While morphologically similar, each species exhibits specific breeding and feeding behaviors. *A. arabiensis,* a close sibling of *A. gambiae,* is found in drier savanna areas of sub-Saharan Africa in association with human populations that combine herding and agriculture. Unlike *A. gambiae,* *A. arabiensis* exhibits an eclectic diet, feeding off of both humans and animals, though it prefers humans. *A. gambiae,* on the other hand, feeds exclusively on humans and prefers moister environments. In all likelihood, this mosquito evolved in association with the expansion of agriculture, but in the absence of alternative animal hosts, in the tropical forests of West and Central Africa. Coluzzi, "The Clay Feet of the Malaria Giant and Its African Roots," 279.

14. Christopher Ehret, "Historical/Linguistic Evidence for Early African Food Production," in J. D. Clark and S. A. Brandt, eds., *From Hunters to Farmers* (Berkeley: University of California Press, 1984): 26–39. Earlier studies have linked the emergence of malaria in the African rain forest with the introduction of iron making and the development of slash-and-burn agriculture, which created environments conducive to *A. gambiae* breeding. The introduction of iron making was dated to 1000 BCE. More recent research, however, indicates that iron age technology is a more recent development within the African rain forest, dating from approximately 2000 years ago. On the other hand, there is linguistic and archaeological evidence that the agricultural development of the rain forest did not require iron making and is in fact much older than previously assumed. Julio Mercader, "Forest People: The Role of African Rainforests in Human Evolution and Dispersal," *Evolutionary Anthropology* 11 (2002): 117–24.

15. Coluzzi, "The Clay Feet of the Malaria Giant and Its African Roots," 280.

16. This would help explain why, as Coluzzi notes, forest dwellers that did not develop slash-and-burn agriculture today possess lower frequencies of the HbS genes.

17. See D. F. C. Clyde, "Recent Trends in the Epidemiology and Control of Malaria," *Epidemiology Reviews* 9 (1987): 219–43, for a discussion of the mechanism by which acquired resistance protects humans from malaria.

18. Carter and Mendis, "Evolutionary and Historical Aspects," 567.

19. Efforts have been made to quantify these different epidemiological conditions in terms of the percentage of young children who are infected with malaria parasites: holoendemic (>75 percent), hyperendemic (50–75 percent), mesoendemic (10–50 percent), hypoendemic (<10 percent).

20. These include forms of hemoglobin polymorphisms with protective values: hemoglobin C, hemoglobin E, and hemoglobin S. See Carter and Mendis, "Evolutionary and Historical Aspects," 570–72, for a discussion of these human genetic mutations.

21. James L. A. Webb, "Malaria and the Peopling of Early Tropical Africa," *Journal of World History* 16, 3 (September 2005): 270–91. Webb argues that the evolution of Duffy negativity occurred over a long period during which human populations penetrated the African tropical forests, combining fishing and agriculture.

22. Carter and Mendis, "Evolutionary and Historical Aspects," 573; F. B. Livingstone, "The Duffy Blood Groups, Vivax Malaria, and Malaria Selection in Human Populations: A Review," *Human Biology* 56, 3 (1984): 413–25.

23. Most recently, a comparison of mitochondrial genome sequences has found a major stepwise growth in the parasite population (a proxy for the increased incidence of human infection) in the period 13,000–8000 BCE. See David J. Conway, "Tracing the Dawn of *Plasmodium falciparum* with Mitochondrial Genome Sequences," *Trends in Genetics* 19, 12 (2003): 671–74, cited in Webb, "Malaria and the Peopling of Early Tropical Africa." This would push the development of the hemoglobin S back to before the second stage of forest agriculture. However, the dating of both these evolutionary developments and of different stages of forest agriculture are open to significant margins of error.

24. Richard Leakey, *Origins* (New York: E. P. Dutton, 1977), 142.

25. Alan R. Templeton, "Out of Africa Again and Again," *Nature* 416 (March 7, 2002): 45–51.

26. It is possible that transmission of *P. malariae* could have occurred in southernmost Europe during warmer interglacial periods. Just how soon after the end of the last ice age malaria vectors and parasites could have expanded into Europe is unclear. De Zulueta argues that it would have taken thousands of years after the end of the last ice age before environmental conditions could have supported this development. J. de Zulueta, "Malaria and Ecosystems: From Prehistory to Posteradication," *Parassitologia* 36 (1994): 7–15. Sallares, on the other hand, points out that recent climatological research indicates that global warming occurred rapidly during what is known as the mid-Holocene climatic optimum years, creating higher average temperatures than existed during the Neolithic and later periods. Sallares, *Malaria and Rome,* 28.

27. While it is theoretically possible that *P. malariae* could have made the long trek through the sub-Artic North to the New World, the thousand-mile trek would have been limited to a single generation in order for the plasmodium to survive. This

seems to be an unlikely scenario, and the movement seems much more likely to have occurred over a longer period of time.

28. Some have argued that *P. malariae* actually had its origins in North America. This idea is based on the existence of a morphologically identical parasite that infects monkeys in South and Central America. Most researchers discount this argument, in part because *P. malariae* existed in southern Europe 2,000 years ago and there was no way for it to have traveled from the New World before the fifteenth century. See Carter and Mendis, "Evolutionary and Historical Aspects," 576; Frederick L. Dunn, "Malaria in the Western Hemisphere," *Human Biology* 37 (1965): 385.

29. Paul Reiter, "From Shakespeare to Defoe: Malaria in England in the Little Ice Age," *Emerging Infectious Disease* 6, 1 (January–February 2000): 5.

30. Carter et al., "Evolutionary and Historical Aspects," 580.

31. Immunological tests identified the histidine-rich antigens of *P. falciparum* in Egyptian mummies dating to 3200 BCE (Sallares, *Malaria and Rome,* 31). Both the methods used in this research and its conclusions, however, have been questioned. See G. M. Taylor et al., "A Sensitive Polymerase Chain Reaction for the Detection of Plasmodium Species DNA in Ancient Human Remains," *Ancient Biomolecules* 1 (1997): 193–203.

32. L. J. Bruce-Chwatt and J. de Zulueta, *The Rise and Fall of Malaria in Europe* (Oxford: Oxford University Press, 1980), 22–23.

33. L. J. Bruce-Chwatt, "Paleogenisis and Paleo-Epidemiology of Primate Malaria," *Bulletin of the World Health Organization* 32 (1965): 376. References to intermittent fevers also appear in the character dictionary *Shuo-wen chieh-tzu,* published during the first century CE, which defines *yao* as "a disease that accompanies intermittent fever" and distinguishes tertian from other types as "a kind of *yao* that occurs every other day." Later references in medical texts beginning in the seventh century distinguish *yao* from a type of malaria prevalent in the South of China called *chang.* Chang was regarded as a more virulent disease caused by poison contained in earth and water of the marshlands in South China. This may suggest that *P. falciparum* malaria was present in the South of China but absent in the North. Kenneth F. Kiple, *The Cambridge World History of Disease* (Cambridge: Cambridge University Press, 1993), 347–48; Miyasita Saburo, "Malaria in Chinese Medicine during the Chin and Yuan periods," 91.

Chapter 2. Malaria Moves North

1. Fiammetta Rocco, *Quinine, Malaria and the Quest for a Cure that Changed the World* (New York: Perrenial, 2003), 36–48.

2. M. D. Grmek, *Diseases in the Ancient Greek World* (Baltimore: Johns Hopkins University Press, 1989).

3. Miriam S. Balmuth, "The Nuraghi of Sardinia: An Introduction," in Miriam S. Balmuth and Robert J. Rowland Jr., *Studies in Sardinian Archeology* (Ann Arbor: University of Michigan Press, 1984), 23–52.

4. Peter Brown, "Malaria in Nuraghic, Punic and Roman Sardinia: Some Hypotheses," in Balmust and Rowland, *Studies in Sardinian Archeology,* 217.

5. Carol Laderman, "Malaria and Progress: Some Historical and Ecological Considerations," *Social Science and Medicine* 9, 11–12 (1975): 587–94.

6. Trapido's biological research indicated that *A. labranchiae*'s existence on the island predated human settlement. H. Trapido, "Biological Considerations in the ERLAAS Project," in J. Logan, *The Sardinian Project* (Baltimore: Johns Hopkins University Press, 1953), 353–75. Bruce-Chwatt and de Zulueta, however, argue that neither *A. labranchiae* nor *A. sacharovi,* could establish themselves on the European coast until the land was cultivated, the soil eroded, and the coast flooded, creating a favorable environment. L. J. Bruce-Chwatt and J. de Zulueta, *The Rise and Fall of Malaria in Europe* (Oxford: Oxford University Press, 1980), 22.

7. Peter Brown, "New Considerations on the Distribution of Malaria Thalassemia and the Glucose-Six-Phosphate Dehydrogenates Deficiency in Sardinia," *Human Biology* 53 (1981): 367–82.

8. Angelo Celli, *The History of Malaria in the Roman Campagna from Ancient Times* (New York: AMS Press, 1977), 16–18.

9. The Etruscan civilization reached its height in the sixth century BCE, but we have little evidence of malaria in central and southern Italy before the fourth century BCE. It is possible that the early development of Etruscan communities occurred before the introduction of malaria, or an efficient vector.

10. Celli, *History of Malaria in the Roman Campagna,* 12–16. Italian malariologist Angelo Celli described the extensive drainage system formed by the construction of tunnels, known as *cuniculi,* under the surface of the Campagna. These tunnels were built between the permeable surface soil and the impermeable layer of compact rock in a strata of semipermeable soil that served to drain the soil of excess water and prevent the formation of areas of saturated soil and marshland that could provide a breeding site for local anopheline mosquitoes. The drainage system appears to have been constructed along the slopes of hills to facilitate cultivation or to provide for the collection of fresh water. Celli noted that at the beginning of the twentieth century three-quarters of the fountains of the Roman Campagna were fed by water from such *cuniculi.* Whether they were intended to open up land for cultivation or as a water collection system is unclear. Either way, the system reduced water runoffs that caused the creation of swampy conditions. See Franco Ravelli and Paula J. Howarth, "Etruscan *cuniculi*: Tunnels for the Collection of Pure Water," Transactions, vol. 2, of the Twelfth International Congress on Irrigation and Drainage, Fort Collins, 1984, organized by the International Commission on Irrigation and Drainage, for a discussion of the history and archaeology of *cuniculi.*

11. Robert Sallares, *Malaria and Rome: A History of Malaria in Ancient Italy* (Oxford: Oxford University Press, 2002), 246.

12. Celli, *History of Malaria in the Roman Campagna,* 19–47; Sallares, *Malaria and Rome,* 252–53.

13. P. A. Brunt, *Italian Manpower, 225 B.C.–A.D. 14* (Oxford: Oxford University Press, 1971); M. S. Spurr, *Arable Cultivation in Roman Italy, c. 200 B.C.– c. A.D. 100* (London: Society for the Promotion of Roman Studies, Journal of Roman Studies Monographs No. 3, 1986). Nathan Rothenstein, *Rome at War: Farms, Families,*

and Death in the Middle Republic (Chapel Hill: University of North Carolina Press, 2004), dates the impact of military service on farming to before the Punic Wars.

14. Sallares, *Malaria and Rome,* 108.

15. Ibid., 109–10; Frank Tenney, *An Economic History of Rome* (Baltimore: Johns Hopkins University Press, 1927), 35. See Gregory S. Aldrete, *Floods of the Tiber in Ancient Rome* (Baltimore: Johns Hopkins University Press, 2007), for a more recent history of flooding in ancient Rome.

16. Sallares, *Malaria and Rome,* 101–2.

17. Bruce-Chwatt and de Zulueta, *The Rise and Fall of Malaria in Europe,* 74–75, 123–24; Peregrine Horden, "Disease, Dragons and Saints: The Management of Epidemics in the Dark Ages," in Terence Ranger and Paul Slack, eds., *Epidemics and Ideas: Essays on the Historical Perception of Pestilence* (Cambridge: Cambridge University Press, 1992), 70–71.

18. Samuel Jeake, *An Astrological Diary of the Seventeenth Century: Samuel Jeake of Rye, 1652–1699,* ed. Michael Hunter and Annabel Gregory (Oxford: Clarendon Press, 1988). Mary Dobson also quotes from Jeake's diary, though she presents a passage from April 1684. That episode, occurring in the spring as opposed to the fall when malaria was most prevalent, probably represented a relapse of a vivax infection. Relapses made malaria a problem over much of the year. Dobson, *Contours of Death and Disease in Early Modern England* (Cambridge: Cambridge University Press, 1997), 314.

19. Reference to a malaria-like sickness in Geoffrey Chaucer's "The Nun's Tale" (1342–1400) have been cited as evidence of the disease's early existence: "You are so very choleric of complexion. / Beware the mounting sun and all dejection, / Nor get yourself with sudden humours hot; / For if you do, I dare well lay a groat / That you shall have the tertian fever's pain, / Or some ague that may well be your bane." Paul Reiter, "From Shakespeare to Defoe: Malaria in England in the Little Ice Age," *Emerging Infectious Disease* 6, 1 (January–February 2000): 3. Yet it is difficult to make a definitive diagnosis using terms popular at the time. While it is possible that malaria had a longer history in England, we do not have evidence of it being a pervasive health problem until the sixteenth century.

20. Dobson, *Contours of Death,* 343.

21. Richard Smith, "Human Resources," in Grenvill Astill and Annie Grant, eds., *The Countryside of Medieval England* (Oxford: Basil Blackwell, 1988), 198.

22. C. G. A. Clay, *Economic Expansion and Social Change: England, 1500–1700,* vol. 1 (Cambridge: Cambridge University Press, 1984).

23. Dobson, *Contours of Death,* 342–50.

24. Keith Lindley, *Fenland Riots and the English Revolution* (London: Heinemann, 1982). The antidraining movement ultimately contributed to the social and economic grievances that helped fuel the English Civil War in the 1640s.

25. H. C. Darby, *The Draining of the Fens* (Cambridge: Cambridge University Press, 1956).

26. Dobson, *Contours of Death,* 342–50.

27. Ibid., 295.

28. Ibid., 25.

29. Jason W, Moore, "The Modern World-System as Environmental History? Ecology and the Rise of Capitalism," *Theory and Society* 32, 3 (June 2003): 307–77.

30. J. R. Wordie, "The Chronology of English Enclosure: A Reply," *Economic History Review,* n.s., 37, 4 (1984): 560.

31. Tom Williamson, "The Disappearance of Malaria from the East Anglia Fens," in Josep L. Barona and Steven Cherry, eds., *Health and Medicine in Rural Europe (1850–1945)* (Valencia: Universitat de Valencia/Libreria, 2005), 249–62.

32. Dobson, *Contours of Death,* 354.

33. Bruce-Chwatt and de Zulueta, *The Rise and Fall of Malaria in Europe,* 74–75.

34. Lewis Hackett, *Malaria in Europe* (Oxford: Oxford University Press, 1937), 95–96.

35. Keith Lindley, *Fenland Riots and the English Revolution* (London: Heinemann, 1982), 2.

36. P. M. Ashburn, *The Ranks of Death: A Medical History of the Conquest of America* (New York: Coward-McCann, 1947), 103–26.

37. Jon Kukla, "Kentish Agues and American Distempers: The Transmission of Malaria from England to Virginia in the Seventeenth Century," *Southern Studies* 25, 2 (1986): 135–47.

38. Mark F. Boyd, *Malariology* (Philadelphia: W. B. Saunders, 1949), 749.

39. Kukla, "Kentish Agues and American Distempers," 135–47.

40. B. R. Carroll, ed., *Historical Collections of South Carolina* (New York: Harper, 1836), 63, cited in Jill Dubisch, "Low Country Fevers: Cultural Adaptations to Malaria in Antebellum South Carolina," *Social Science and Medicine* 21, 6 (1985): 643.

41. It is impossible to know with any certainty whether the same mosquitoes were present during the seventeenth and eighteenth centuries, or whether their breeding and feeding habits were the same as they were when these twentieth-century studies were conducted. But *A. quadrimaculatus* occurs throughout the eastern United States and breeds in similar environments, albeit with some variation, so the vector was probably a longtime resident of South Carolina and exhibited similar breeding preferences during earlier periods.

42. The normal flight range of *A. quadrimaculatus* is one mile. J. A. Prince and T. H. D. Griffitts, "Flight of Mosquitoes. Studies of the Distance of Flight of *Anopheles quadrimaculatus,*" *U.S. Public Health Reports* 32, 18 (May, 1917): 656–59. On the other hand, a 1945 study found that a few mosquitoes breeding in swamps located in the center of the Santee reservoir in South Carolina traveled up to three miles.

43. M. F. Boyd and W. K. Startman-Thomas, "The Comparative Susceptibility of Anopheles quadrimaculatus Say, and Anopheles crucians Weid. (Inland Variety) to Parasites of Human Malaria," *American Journal of Hygiene* 20, 1 (July 1934): 247–57. Moreover, in contrast to *A. quadrimaculatus, A. crucians* prefers to feed out of doors. A study conducted in the Tidewater region of Virginia suggested that there may be two species of *A. crucians,* one that prefers brackish water, the other preferring fresh water. The brackish variety appeared to be a more efficient transmitter of malaria.

44. *A. crucians* requires high densities of gametocytes in the host population.

When the gametocyte densities are low, *A. crucians* are less likely to become infected and thus to transmit malaria.

45. St. Julien Ravenel Childs, *Malaria and Colonization in the Low Country, 1526–1696* (Baltimore: Johns Hopkins Press, 1940), 130–32.

46. Joyce E. Chaplin, *An Anxious Pursuit: Agricultural Innovation and Modernity in the Lower South, 1730–1815* (Chapel Hill: North Carolina Press, 1993), 228.

47. H. Roy Meterns and George D. Terry, "Dying in Paradise: Malaria, Mortality, and the Perceptual Environment in Colonial South Carolina," *Journal of Southern History* 50, 4 (1984): 547.

48. African slaves may have contributed knowledge of rice growing as well as labor. Peter Wood, *Black Majority: Negroes in Colonial South Carolina from 1670 through the Stono Rebellion* (New York: Alfred A. Knopf, 1975), 58–60. Judith Carney took up this theory and developed it in "Landscapes of Technology Transfer: Rice Cultivation and African Continuities," *Technology and Culture* 37, 1 (1996): 5–35.

49. Wood, *Black Majority*, 151–52.

50. George David Terry, "'Champaign Country': A Social History of an Eighteenth Century Lowcountry Parish in South Carolina, St. Johns Berkeley Country" (Ph.D. dissertation, University of South Carolina, 1981), 92–63.

51. The belief may have been true since, as we have seen in chapter 1, the presence of both Duffy negativity and hemoglobin S mutations among large numbers of Africans in Central and West Africa from which most slaves were drawn made these slave populations less susceptible to malaria infection in South Carolina. Phillip Curtin, "The Epidemiology of the Atlantic Slave Trade," *Political Science Quarterly* 83 (1968): 216.

52. Merrens and Terry, "Dying in Paradise," 546–48. Mark F. Boyd, "An Historical Sketch of the Prevalence of Malaria in North America," *American Journal of Tropical Medicine* 21 (1941): 228.

53. Chaplin, *An Anxious Pursuit*, 232–37.

54. Boyd, "An Historical Sketch of the Prevalence of Malaria in North America," 236–37.

55. Boyd, *Malariology*, 750.

56. John C. Hudson, *The Making of the Corn Belt: A Geographical History of Midwestern Agriculture* (Bloomington: Indiana University Press, 1994), 3.

57. Douglas K. Meyer, *Making the Heartland Quilt: A Geographical History of Settlement and Migration in Early-Nineteenth-Century Illinois* (Carbondale: Southern Illinois University Press, 2000), 14–25.

58. V. C. Vaughan, *Epidemiology and Public Health*, vol. 2 (St. Louis, 1923), 594, cited in Erwin H. Acherknecht, *Malaria in the Upper Mississippi Valley, 1760–1900* (Baltimore: Johns Hopkins Press, 1945), 28. Despite the presence of malaria, which contributed to Illinois's early reputation as "a grave yard," the availability of fertile tracts of land continued to attract new settlers. As a consequence, the population of Illinois country increased from about 15,000 in 1815 to almost 40,000 by the end of 1818. By the end of the 1830s the population had expanded to nearly 157,000, as an increasing number of new settlers were determined to develop productive market-oriented farms.

59. Meyer, *Making the Heartland Quilt,* 100–102.

60. Hudson, *The Making of the Corn Belt,* 96–99.

61. John M. Peck, *A New Guide for Immigrants to the West* (Boston: Gould, Kendal and Lincoln, 1936), 115, cited in Meyer, *Making the Heartland Quilt,* 31.

62. The state's leading expert on the disease asserted that California in the 1910s was the most malarious state in the country with between 5,000 and 6,000 cases a year. Linda Nash, *Inescapable Ecologies: A History of Environment, Disease, and Knowledge* (Berkeley: University of California Press, 2006).

63. Curtis R. Best. "A History of Mosquitoes in Massachusetts," paper presented at the annual meeting of the Northeast Mosquito Control Association, Salem, Massachusetts, November 1993.

Chapter 3. A Southern Disease

1. Margaret Humphreys, *Malaria, Poverty, Race and Public Health in the United States* (Baltimore: Johns Hopkins University Press, 2001), 37.

2. M. A. Barber, "The History of Malaria in the United States," *Public Health Reports* 44, 43 (1929): 2579.

3. Lee J. Alston, "Issues in Postbellum Southern Agriculture," in Lou Ferleger, ed., *Agriculture and National Development: Views on the Nineteenth Century* (Ames: Iowa State University Press, 1990), 207–28; Dwight B. Billings Jr., *Planters and the Making of a "New South": Class, Politics, and Development in North Carolina, 1865–1900* (Chapel Hill: University of North Carolina Press, 1979).

4. Black farmers tried to negotiate contracts that would ensure them as much independence as possible, preferring various forms of rent tenancy. Planters, for their part, often preferred labor tenancy, or wage contracts. During the last two decades of the nineteenth century, many sharecropping and tenancy contracts turned into forms of wage labor.

5. In Mississippi, Georgia, and Florida, state laws labeled blacks not under contract to a planter, or otherwise employed, as vagrants and subject to arrest. Other informal agreements and anti-enticement laws restricted the movement of free blacks and helped keep them trapped in tenancy contracts. Stephen J. DeCanio, *Agriculture in the Postbellum South: The Economics of Production and Supply* (Cambridge, Mass.: MIT Press, 1974), 18.

6. In the newly opened cotton areas of Mississippi, Arkansas, and Texas, landowners were less able to capture black labor as labor shortages created competition for workers. In this situation, blacks frequently moved from plantation to plantation, hoping to improve their status.

7. The debt burden was further increased by planters' insistence that tenants devote all their land to cotton production. This meant that tenants depended on income from cotton to purchase food as well as other necessities, because they often could not grow enough food for themselves. In a bad year, when the crop was small due to poor weather or a blight of boll weevils, or when the price of cotton fell, tenants were forced to take on more debt to feed their families. DeCanio, *Agriculture in the Postbellum South,* 48. See also Harold D. Woodman, *New South, New Law: The*

Legal Foundations of Credit and Labor Relations in the Postbellum Agricultural South. (Baton Rouge: Louisiana State University Press, 1995).

8. Gilbert C. Fite, *Cotton Fields No More: Southern Agriculture, 1865–1980* (Lexington: University Press of Kentucky, 1984).

9. G. W. Herrick, "The Relation of Malaria to Agriculture and Other Industries in the South," *Pop Science Monthly,* April 1903, 522.

10. W. B. Brierly examined malaria in relationship to income and cotton prices in Mississippi between 1916 and 1937. W. B. Brierly, "Malaria and Socio-Economic Conditions in Mississippi," *Social Forces* 4 (1945): 451–59.

11. Barber, "The History of Malaria in the United States."

12. Humphreys, *Malaria, Poverty, Race,* 54.

13. M. A. Barber and W. H. W. Komp, "Malaria in the Prairie Rice Regions of Louisiana and Arkansas," *Public Health Reports* 41, 45 (1926): 2527–49. He noted that a single building could contain thousands of mosquitoes, and surfaces were sometimes blackened by large numbers of roosting Anopheles.

14. Pete Daniel, *Breaking the Land: The Transformation of Cotton, Tobacco, and Rice Cultures since 1880* (Urbana: University of Illinois Press, 1985).

15. G. C. Fite, *Cotton Fields No More: Southern Agriculture, 1865–1980* (Lexington: University Press of Kentucky, 1984), 11; Daniel, *Breaking the Land,* xii–xiii.

16. P. B. Haney et al., "Cotton Production and the Boll Weevil on Georgia: History, Costs of Control and Eradication," College of Agriculture and Environmental Sciences, University of Georgia, *Research Bulletin* 428 (1996): 9–12.

17. Stewart Tolnay and E. M. Beck have argued that the high tide of black emigration from the South occurred during the 1920s and was fueled by a dramatic rise in racial violence as much as by declining economic opportunities. "Black Flight: Lethal Violence and the Great Migration, 1900 to 1930," *Social Science History* 19, 3 (Autumn 1990): 347–70.

18. Humphreys, *Malaria, Poverty, Race,* 110–12.

19. Frank Snowden, "Mosquitoes, Quinine and the Socialism of Italian Women," *Past and Present* 178, 1 (February 2003): 176–209.

20. Frank M. Snowden, *Violence and the Great Estates in the South of Italy* (Cambridge: Cambridge University Press, 1986), 8–9.

21. Ibid., 15.

22. Frank Snowden, *The Conquest of Malaria: Italy, 1900–1962* (New Haven: Yale University Press, 2006), 18.

23. In contrast to England, the concentration of lands associated with the capitalization of agriculture was not accompanied by the growth of other industries, and the population of workers continued to exceed demand.

24. Frank Snowden, "'Fields of Death': Malaria in Italy, 1861–1962," *Modern Italy* 4, 1 (1999): 35.

25. Snowden, "Mosquitoes, Quinine and the Socialism of Italian Women," 176–209.

26. Manio Rossi-Doria, "The Land Tenure System and Class in Southern Italy," *American Historical Review* 64, 1 (october 1958): 46–53.

27. Snowden, *The Conquest of Malaria,* 84–85.

Chapter 4. Tropical Development and Malaria

1. See Michael Watts, *Silent Violence: Food, Famine and Peasantry in Northern Nigeria* (Berkeley: University of California Press, 1983), for a detailed discussion of the potential food security costs of African peasant involvement in cash-crop production. It should be noted that, while peasants bore the brunt of declining commodity prices, they seldom benefited from price increases. Marketing boards seldom passed windfall profits on to producers.

2. Mining was a notable exception to the absence of industry. However, as a series of studies have shown, employment in mining seldom provided workers with resources needed to substantially improve their standard of living.

3. R. K. Newman, "Social Factors in the Recruitment of the Bombay Mill-hands," *Economy and Society: Essays in Indian Economic and Social History* (Delhi: Oxford University Press, 1979), 277–95.

4. *Report of the Royal Commission on Labour in India* (London: Government Stationary Office, 1931).

5. Vinay Kamat, "Resurgence of Malaria in Bombay (Mumbai) in the 1990s: A Historical Perspective," *Parassitologia* 42, 1–2 (2000): 135–48.

6. It now appears that the link between malaria and irrigation in Africa, like much else in the epidemiology of this disease, was mediated by a complex set of biological, social, and economic factors. J. N. Ijumba, R. W. Mwangi, et al., "Malaria Transmission Potential of Anopheles Mosquitoes in the Mwea-Tebere Irrigation Scheme, Kenya," *Medical and Veterinary Entomology* 4, 4 (1990): 425–32. As Tamara Giles-Vernick has recently noted, a large number of variables determined whether irrigation farming increased the incidence of malaria, including the distance that farmers lived from the irrigated fields, cultivation methods, the degree to which specific mosquitoes preferred humans, farmers' living conditions, the presence of alternative sources of blood meals for mosquitoes (including cattle), available health services, and changing gender relations. T. Giles-Varneck, "Malaria, Mosquitoes, and Environmental Change: Rethinking Etiologies of an Epidemic in French Soudan, 1906–1938," Medicine in Africa, Bryn Mawr College/Haverford College (2005). Also important was the extent to which the farming population had been exposed to malaria before irrigation was begun. Thus it has been argued that irrigation was more likely to increase the incidence of malaria in areas of unstable transmission, where people had little or no immunity to malaria parasites, such as the African highlands and formerly arid environments. But for areas where malaria was already stable, the introduction of crop irrigation had little impact on malaria transmission. Indeed, there is recent evidence that in some sites there is less malaria in irrigated communities than surrounding areas. The explanation for this finding is still unresolved, but research from rice irrigation farming areas in western Kenya suggests that in some cases, irrigation may lead to the displacement of a highly endophilic and anthropophilic malaria vector with one that has a lower vectorial capacity. In the Kenya case, *A. funestus* Giles was pushed out by *A. arabiensis* Patton mosquitoes. The numbers of *A. arabiensis* increased dramatically in irrigation sites, but because this vector fed on cattle as well as humans, transmission declined. M. C. Mutero, C. Kabutha, et al.,

"A Transdisciplinary Perspective on the Links between Malaria and Agroecosystems in Kenya," *Actica Tropical* 89, 2 (2004): 171–86.

7. J. Schwetz, *Recherches sur le paludisme dans la bordure orientale du Congo Belge*, Institut Royal Colonial Belge Memoires, vol. 14, 3 (Brussells, 1944).

8. Alan H. Jeeves, "Migrant Workers and Epidemic Malaria on the South African Sugar Estates, 1906–1948," in Alan H. Jeeves and Jonathan Crush, eds., *White Farms, Black Labor: The State and Agrarian Change in Southern Africa, 1910–50* (Oxford: James Currey, 1997), 114–36.

9. Ralph Shiomowitz and Brennan Lance, "Mortality and Indian Labor in Malaya," *Indian Economic and Social History Review* 29, 1 (1992): 57–75.

10. Schwetz, *Recherches sur le paludisme*. Similarly, gold mines established in the Ituri forest by the Belgians recruited labor from the highland areas of Kivu district, exposing these workers to malaria for the first time.

11. Ivo Mueller et al., "Epidemic Malaria in the Highlands of Papua New Guinea," *American Journal of Tropical Hygiene* 72, 5 (2005): 554–60.

12. J. Mouchet, S. Laventure, et al., "The Reconquest of the Madagascar Highlands by Malaria," *Bulletin de la Société de Pathologie Exotique* (Paris) 90, 3 (1997): 162–68; S. W. Lindsay and W. J. M. Martens, "Malaria in the African Highlands: Past, Present and Future," *Bulletin of the World Health Organization* 76, 1 (1998): 33–45.

13. The following discussion is drawn from R. M. Packard and P. Gadhela, "A Land Filled with Mosquitoes: Fred L. Soper, the Rockefeller Foundation, and the *Anopheles gambiae* Invasion of Brazil," *Parassitologia* 36 (1994): 197–213.

14. Information on the ecology and history of northeast Brazil is drawn largely from Manuel Correia de Andrande, *Land and People of Northeast Brazil* (Albuquerque: University of New Mexico Press, 1980). Also consulted, Billy Jaynes Chandler, *The Feitosas and the Sertão dos Inhamuns: The History of a Family and Community in Northeast Brazil, 1700–1930* (Gainsville: University of Florida Press, 1972); E. B. Resnick, *The Peasant in the Sertão: A Short Exploration of His Past and Present* (Leiden: Institute of Cultural and Social Studies, Leiden University, 1981); Josue de Castro, *Death in the Northeast* (New York: Random House, 1966); Jason W. Clay, "The Articulation of Non-Capitalist Agricultural Production Systems with Capitalist Exchange Systems: The Case of Garanhuns, Brazil" (Ph.D. dissertation, Cornell University, 1979).

15. Cattle raising, like agriculture, was conducted through a system of sharing. The land owner hired "cowboys" to herd his cattle. In return the cowboy received one-fourth of the calves produced from the herd he tended. The cowboy then sold these to the landowner.

16. Andrande, *Land and People of Northeast Brazil*, 229. During 1936 and 1937, the average day wage was three cruzeiros. By comparison the cost of provisions was high. During the late 1930s inflation limited this accumulation still further by increasing the price of subsistence goods. Quoting the prices paid for common provisions purchased by migrants in Alagoas, Andrade noted that dried codfish sold for Cr3.50 per kilo, coffee for 2.80, sugar for 1.00, butter for 9.50. Manioc sold for 0.80 per liter and beans for 1.40.

17. Ibid., 171.

18. Ibid., 20.

19. Rockefeller Archives Center, Sleepy Hollow, New York, Annual Report, 1939, p. 15, RG5, series 3, box 113.

20. *Gazeta de Noticias, Fortaleza of April 11, 1940,* cited in Dr. Gerry F. Killeen et al., "Eradication of *Anopheles gambiae* from Brazil: Lessons for Malaria Control in Africa?" *Lancet Infectious Diseases* 10 (October 2, 2002): 620.

21. S. R. Christophers, *Malaria in the Punjab* (Calcutta: Superintendent Government Printing, 1911).

22. S. Zurbrigg, "Re-thinking the 'Human Factor' in Malaria Mortality: The Case of Punjab, 1868–1940," *Parassitologia* 36, 1–2 (1994): 133.

23. A. K. Sen, *Poverty and Famines: An Essay on Entitlement and Deprivation* (New York: Clarendon Press, 1981).

24. Aktar Husain Siddiqi, "Nineteenth Century Agricultural Development in the Punjab," *Indian Economic and Social History Review* 21, 3 (1984): 293–312; Imran Ali, "Malign Growth, Agricultural Colonization and the Roots of Backwardness in the Punjab," *Past and Present* 114 (1987): 110–32.

25. Dipak Chattaraj, "Curse amidst Blessing: Canal Irrigation in Colonial South-Eastern Punjab, 1858–1919," *Indian Historical Review* 27, 2 (2000): 122–46; Indu Agnihorti, "Ecology, Land Use and Colonization in the Canal Colonies of Punjab," *Indian Economic and Social Review* 33, 1 (1996): 37–59.

26. Malaria may have occasionally spread into the highlands during the nineteenth century. Periods of drought and famine occasionally led people living in lowland malarious areas to look for food in neighboring highlands where malaria was uncommon. Such movements could contribute to temporary expansions of malaria into highland areas. It is also possible that transformations in patterns of agricultural production associated with the early development of international commodity markets during the nineteenth century may have altered transmission patterns in tropical areas of West Africa.

27. See note 6.

28. European researchers working in the lowveld in the 1930s observed a marked decrease in parasite rates among Africans as one moved from the lowest areas of the lowveld to higher locations. Randall M. Packard, "'Malaria Blocks Development' Revisited: The Role of Disease in the History of Agricultural Development in the Eastern and Northern Transvaal Lowveld, 1890–1960," *Journal of Southern African Studies* 27, 3 (2001): 591–612.

29. Report of the Carnegie Commission, *The Poor White Problem in South Africa,* part 4: *Health Factors in the Poor White Problem* (Stellenbosch, 1932), 104.

30. Ibid., 89–90.

31. H. A. Spencer, "Epidemic Malaria," *Transvaal Medical Journal,* February 1910, 138.

32. It is difficult to know whether the high mortality experienced by Africans during this regional epidemic reflected the virulence of the infection, or the lack of immunity among highland workers, or a combination of the two. Interestingly, those infected with the disease believed that it was new, brought in by workers who

were recruited to work on the railway from Mozambique (the local designation of the disease was "Mashona sickness," as Mashona was the name given to workers coming from Mozambique). We know that early European explorers further north in Mozambique and Zimbabwe suffered severe bouts of fever in the sixteenth century. However, the absence of the sickle cell trait among African populations in South Africa suggests that either falciparum malaria was not present at earlier times or transmission was not sufficiently intense to create selective pressure for sickle cell development. It is therefore possible that falciparum malaria was introduced further south relatively recently, timing that may help to account for the movement of Africans out of the lowlands at the end of the nineteenth century.

33. See Removal of Natives from Farm New Hanover 750 Zoutspansberg 1909–17, NA 139 849/1909/F263; Barberton Native Area, Beaumont and Eastern Transvaal Land Commission and Released Native Areas (Native Land Act 1913 amended) 1928–52; LDE 912 18318; Native Land Commission, Recommendations of Farms in Zouspansberg Reserved for Natives, LDE 912 18313, for discussion of African removals; all in Central Archives Depot, Pretoria.

34. Based on a comparison of South African Census data from 1921 and 1936 for lowveld districts.

35. Various Correspondence located in file "Malaria Fever Transvaal 1913–1930," NTS 6718 30/315, Central Archives Depot, Pretoria.

36. Magistrate Zoutpansberg to Sec for Public Health, April 6, 1936, GES 2648 24/56; "Comments Re: Screening of Native Huts," August 3, 1932, GES 2615 2/56L; both in Central Archives Depot, Pretoria.

37. Packard, "'Malaria Blocks Development' Revisited," 591–612.

38. For a history of malaria epidemics in colonial Swaziland, see R. M. Packard, "Maize, Cattle, and Mosquitoes: The Political Economy of Malaria Epidemics in Colonial Swaziland," *Journal of African History* 25, 2 (1984): 189–212.

39. Annual Report, Sister R. C. de Villiers, Health Inspector, Waterberg, July 1939–June 1940, GES 1913 38/32A, Central Archives Depot, Pretoria.

40. Report of Malaria Control Station at Tzaneen for Period 1/7/32 to 30/6/33, GES 1913 38/32A, Central Archives Depot, Pretoria. The role of dams in the spread of malaria was recognized as early as 1930 by entomologist Botha de Mellion. He wrote, "Attention is drawn to the present construction of small dams next to irrigation furrows for the purpose of conserving water. The nature of the soil and the method of construction make their complete drainage impossible. We believe that the small pools resulting from incomplete drainage will be a great source of *A. gambiae* when the vector is present. . . . The above opinion is based on a hypothesis supported by very meager evidence, it must be admitted, that *A. gambiae* generally dies out in the winter in many areas where malaria is not a yearly occurrence. Then soon as conditions become favorable again these mosquitoes advance from their permanent breeding grounds into areas from which they are normally absent. Naturally this advance is helped by the amount of open water available. If this is so then a series of dams of the above mentioned description stretching from the banks of the Crocodile river up to Brits would form an ideal road of advance for *A. gambiae*." Botha de Meillon,

Report of the Entomologist on a Tour of the Transvaal Lasting from 14 April to 6 May 1930, GES 2614 2/56H, p. 12, Central Archives Depot, Pretoria.

Chapter 5. The Making of a Vector-Borne Disease

1. Angelo Celli, *Malaria: According to the New Researches* (New York: Longmans, Green, 1900), 2–3.

2. Ibid., 253.

3. M. A. Barber and W. H. W. Komp, "Malaria in the Prairie Rice Regions of Louisiana and Arkansas," *Public Health Reports* 41, 45 (1926): 2546.

4. G. Bodeker, M. Willcox, et al., *The Therapeutic Potential of Plants Used Traditionally as Antimalarials: New Directions for Research* (1998): 287–89; R. Caniato and L. Puricelli, "Review: Natural Antimalarial Agents (1995–2001)," *Critical Reviews in Plant Sciences* 22, 1 (2003): 79–105; S. Tagboto and S. Townson, "Antiparasitic Properties of Medicinal Plants and Other Naturally Occurring Products," *Advances in Parasitology* 50 (2001): 199–295.

5. Lewis Hackett, *Malaria in Europe: An Ecological Study* (Oxford: Oxford University Press, 1937), 18.

6. See Gordon Harrison, *Mosquitoes, Malaria and Man: A History of the Hostilities since 1880* (London: John Murray, 1978), for a discussion of these developments.

7. Elizabeth Fee, *Disease and Discovery: A History of the Johns Hopkins School of Hygiene and Public Health, 1916–1939* (Baltimore: Johns Hopkins University Press, 1987), 18–19.

8. Anne-Marie Moulin, "The Pasteur Institute between the Two Wars," in Paul Weindling, ed., *International Health Organizations and Movements, 1918–1939* (Cambridge: Cambridge University Press, 1995), 253.

9. Paul Weindling, "Social Medicine at the League of Nations Health Organization and the League of Nations," in ibid., 138–41.

10. Robert Desowitz, *Malaria Capers: More Tales of Parasites and People, Research and Realities* (New York: Norton, 1991), 199.

11. William Bynum, "An Experiment That Failed: Malaria Control at Mian Mir," *Parassitologia* 36, 1–2 (1994): 107–20. Sheldon Watts has argued that resistance to vector control in India was also related to the miles of irrigation canals built by the British over wide areas of northern India to encourage agricultural production. Any serious consideration of vector control would implicate these canals in promoting malaria (Watts, "British Development Policies and Malaria, 1897–1929," *Past and Present* 165 [1999]: 141–81). As we saw in the preceding chapter, this implication was not unreasonable.

12. William Crawford Gorgas, *Sanitation in Panama* (New York: D. Appelton and Company, 1918), 182–205.

13. "History of Malaria, an Ancient Disease," CDC Malaria Web site, www.cdc.gov/malaria/history/index.htm.

14. See Frank Snowden, *The Conquest of Malaria: Italy, 1900–1962* (New Haven: Yale University Press, 2006), for a discussion of these efforts.

15. See Hugh Evans, "European Malaria Policies in the 1920s and 30s," *ISIS* 80

(1989): 40–59, for a fuller discussion of competing theories of malaria control in Europe.

16. M. Harrison, "Medicine and the Culture of Command: The Case of Malaria Control in the British Army during the Two World Wars," *Medical History* 40, 4 (1996): 437–52.

17. Snowden, *The Conquest of Malaria*, 82–83.

18. Health Organization, Malaria Commission, *Report on Its Tour of Investigation in Certain European Countries, 1924* (Geneva: Publication Department of the League of Nations, 1925), 63.

19. Health Organization, Malaria Commission, *Principles and Measures of Antimalarial Measures in Europe* (Geneva: Publication Department of the League of Nations, 1927), 9.

20. S. P. James, "Remarks on Malaria in Kenya," *Kenya and East African Medical Journal* 6, 4 (July 1929): 96.

21. Health Organization, Malaria Commission, *Report on Its Tour of Investigation in Certain European Countries, 1924*, 161. The report notes that malaria mortality dropped from 2,523, between 1882 and 1901 to 482 between 1902 and 1921.

22. Frank Snowden, " 'Fields of Death': Malaria in Italy, 1861–1962," *Modern Italy* 4, 1 (1999): 25–57.

23. Snowden, *The Conquest of Malaria*, 146.

24. R. Neveu, "L'assainissement des Marais-Pontins," *Annales d' Hygiene Publique Industustrielle et Sociale* 9, 2 (February 1931): 81–86.

25. Snowden, *The Conquest of Malaria*, 142–81.

26. Uriel Kitron, "Malaria, Agriculture and Development: Lessons from Past Campaigns," *International Journal of Health Services* 17, 2 (1987): 295–326. See also Andrew Spielman and Richard DiAntonio, *Mosquito: A Natural History of Our Persistent and Deadly Foe* (New York: Hyperion, 2001), 131, 152–53.

27. Ibid. TVA data show that malaria, as measured by the percentage of people who tested positive for parasites, dropped from just under 30 percent in 1933 to nearly 0 percent in 1944. See also Humphreys, *Malaria, Poverty, Race*, 103–5.

28. Similar integrated approaches to malaria control were used in Mexico. The Papaloapan Commission in the state of Veracruz transformed the Papaloapan River Valley project, providing public water deposits, latrines, antiparasitic clinics, and hygiene education programs alongside extensive agricultural development and the provision of electricity. H. Gomez-Dantes and A.-E. Birn, "Malaria and Social Movements in Mexico: The Last 60 Years," *Parassitologia* 42 (2000): 69–85.

29. Eric D. Carter, "Disease, Science and Regional Development: Malaria Control in Northwest Argentina" (Ph.D. dissertation, University of Wisconsin–Madison, 2005); C. A. Alvarado, "Measures against Malaria in Argentina," *Bolletin sanitaria dept nacionale Higene Argentine* 2, 5 (May 1938): 451–63.

30. Weindling, "Social Medicine at the League of Nations Health Organization," 144–46.

31. A. Štampar, "Observations of a Rural Health Worker," *New England Journal of Medicine* 218, 24 (1938): 994.

32. See Socrates Litsios, "Selskar Gunn and China: The Rockefeller Foundation's

'Other' Approach to Public Health," *Bulletin of the History of Medicine* 79, 2 (2005): 295–318.

33. Socrates Litsios, *The Tomorrow of Malaria* (Wellington: Pacific Press, 1996), 133. The conference also made two other novel recommendations. One was that women had a key role to play in the development of rural hygiene. The second was that programs should not be imposed from above but should involve local participants in planning activities.

34. A. Missiroli, L. W. Hackett, et al., "The Races of *A. maculipennis* and Their Importance in the Distribution of Malaria in Some Regions of Europe," *Rivista di Malariologia* 12, 1 (1933): 1–56.

35. This success was repeated in the early 1940s in Egypt. There *A. gambiae* had moved northward from the Sudan, producing major outbreaks of malaria in upper Egypt and threatening to move into lower Egypt where the British army was preparing to defend the Suez Canal from German advances. Soper once again organized a successful anti-*A. gambiae* effort. This episode occurred in wartime and, in fact, was cloaked in secrecy. As such the foundation was unable to use it to push its support for vector control.

36. Rockefeller Archives Center, Sleepy Hollow, New York, F. Soper, "*Anopheles gambiae* 1941," memo enclosed in letter to R. B. Fosdick, January 21, 1942, RF RG 1.1, series 305I, box 16, folder 141.

37. For a recent interesting discussion of this episode, see Timothy Mitchell's *Rule of Experts: Egypt, Techno-politics, Modernity* (Berkeley: University of California Press, 2002).

38. R. J. T. Joy, "Malaria in American Troops in the South and Southwest Pacific in World War II," *Medical History* 43, 2 (1999): 192–207.

39. Leo Slater, "Malaria Chemotherapy and the 'Kaleidoscopic' Organization of Biomedical Research during World War II," *AMBIX* 51, 2 (July 2004): 107–34.

40. John H. Perkins, "Reshaping Technology in Wartime: The Effect of Military Goals on Entomological Research and Insect Control Practices," *Technology and Culture* 19, 2 (1978): 175.

41. Thus during the 1940s economic entomologists, who represented the branch of entomology concerned with its practical applications, became committed to chemical control at the expense of other methods and, according to some entomologists, "at the expense of the scientific discipline itself." As early as 1943, Percy Annand, in his presidential address to the fifty-fifth annual meeting of the American Association of Economic Entomologists, noted that research efforts of long duration on the biology of insect pests had been disrupted and changed to the search for immediate solutions for short-term military purposes. The publication of articles in disciplinary journals reflected this shift in research priorities. The percentage of papers in the *Journal of Economic Entomology* on general biology of insect pests and their biological control dropped from 33 percent in 1937 to 17 percent in 1947. Conversely, the percentage devoted to testing of insecticides rose from 58 to 76 percent. Not only were biological methods of control replaced or disrupted by the use of chemicals, but entomologists who persisted in attempting to develop control procedures using both biological and chemicals were "ridiculed by the dominating

chemical control proponents as a lunatic fringe of economic entomologists." Perkins, "Reshaping Technology in Wartime," 183–84.

42. Ibid., 182.

43. Robert van den Bosch, *The Pesticide Conspiracy* (Garden City, N.Y.: Doubleday, 1978), 21.

44. This act of sabotage by the retreating Germans was intended to reintroduce malaria to the region, both to punish the Italians for their withdrawal from the war and to serve as an impediment to the Allied advance. Frank Snowden has argued that this was an early example of biological warfare. Snowden, *The Conquest of Malaria*.

45. Rockefeller Archives Center, Sleepy Hollow, New York, Paul Russell's Diaries, 1947–48, September 19, 1947, Meeting of the Board of Scientific Directors of the International Health Division at the Rockefeller Institute in New York.

46. Snowden, *The Conquest of Malaria*, 206–7.

47. Expert Committee on Malaria, Report on Second Session, May 19–25, 1948, Washington, D.C., WHO.I/205, June 8, 1948, p. 49.

48. Paul Russell's Diaries, Rockefeller Foundation Archives, May 25, 1948; Malaria Expert Committee Report, 69.

49. WHO, 5th Expert Committee on Malaria Meeting, Istanbul, September 7–12, 1953, p. 6.

50. Martin David Dubin, "The League of Nations Health Organization," in Paul Weindling, ed., *International Health Organization and Movements, 1918–1939* (Cambridge: Cambridge University Press, 1995), 68–71.

51. Harry Cleaver, "Malaria and the Political Economy of Public Health," *International Journal of Health Services* 7, 4 (1977): 557–79, has argued that these interests were the primary motivating factors behind the Eisenhower administration's later support for the WHO global eradication program in 1957. Yet it is evident from United States Overseas Mission records from Vietnam and Thailand that even in the early fifties, malaria control programs were very much part of the U.S. war against communism. See R. Packard, "Malaria Dreams: Postwar Visions of Health and Development in the Third World," in *Malaria and Development,* special issue of *Medical Anthropology,* guest editor, Randall M. Packard, 17, 3 (1997): 279–94.

52. A Joint Statement on Public Health Priorities issued by Public Health Division of the Foreign Operations Administration and the Public Health Service and Children's Bureau of the U.S. Department of Health, Education and Welfare in March 1954 provides the following first principles upon which technical assistance health programs will be supported: "Strengthen economy by health benefits which release effective human energy, improve citizen morale, improve environment for local and foreign investment, open new land and project areas; contribute to our political objectives by reaching large populations with highly welcomed personal service programs, by demonstrating our deep human interest in man and his dignity." U.S. National Archives, Washington, D.C., RG469, Office of Public Services, Public Health Division, 1952–59.

53. International Development Advisory Board, Report and Recommendations on Malaria Eradication, April 13, 1956, p. 14 [Vector Borne Control Archives, Washington, D.C. 008195].

54. L. J. Bruce-Chwatt, "The Challenge of Malaria: Crossroads or Impasse?" in Clive Wood, *Tropical Medicine from Romance to Reality* (New York: Grune and Stratton, 1978), 15.

55. International Development Advisory Board, Report and Recommendations on Malaria Eradication, April 13, 1956, p. 11 [VBC 008195].

56. Ibid., p. 8.

57. William Vogt, *Road to Survival* (New York: William Sloane Associated, 1948), 13. I wish to thank Socrates Litsios for bringing this book to my attention.

58. Ibid., 28.

59. Gunnar Myrdal, "Economic Aspects of Health," *WHO Chronicle* 6 (1952): 203–18.

60. C.-E. A. Winslow, *The Cost of Sickness, Price of Health,* Monograph Series No. 7 (Geneva: World Health Organization, 1951): 81.

61. Paul F. Russell, *Man's Mastery of Malaria* (London: Oxford University Press, 1955), 257.

62. Socrates Litsios, "Malaria Control, the Cold War and the Postwar Reorganization of International Assistance," *Medical Anthropology* 17, 3 (1997): 255–78.

Chapter 6. Malaria Dreams

1. George Macdonald, *The Epidemiology and Control of Malaria* (Oxford: Oxford University Press, 1957); Amar Hamoudi and Jeffrey Sachs, "The Changing Global Distribution of Malaria: A Review," Harvard Center for International Development Working Paper No. 2 (1999).

2. The following discussion of the launching of the Malaria Eradication Programme is drawn largely from my earlier article: " 'No Other Logical Choice': Global Malaria Eradication and the Politics of International Health," *Parassitologia* 40, 1–2 (June 1998): 217–30.

3. World Health Organization, "Malaria Eradication, Proposal by the Director-General," A8/P&B/10, May 3, 1955, 9.

4. Rockefeller Archives Center, Sleepy Hollow, New York, Paul Russell's Diaries, January 8, 1948, Washington, D.C., NIH Malaria Conference.

5. Rockefeller Archives Center, Sleepy Hollow, New York, Paul Russell's Diaries, September 27, 1948, Washington, D.C., NIH Malaria Study Section meetings.

6. Dr. Fred L. Soper, Pan American Sanitary Bureau, "Hemisphere-Wide Malaria Eradication," 28, statement Washington D.C., February 14, 1955.

7. Personal communication from J. Najera, Geneva, June 8, 1992.

8. For a discussion of these efforts, see Vance Packard, *The Hidden Persuaders* (Boston: David McKay, 1957).

9. Department of State, *Point Four: Cooperative Program for AID in the Development of Underdeveloped Areas,* Department of State Publication 3719, Economic Cooperation Series (Washington, D.C.: U.S. Printing Office, 1949), 24.

10. H. Gomez-Dantes and A.-E. Birn, "Malaria and Social Movements in Mexico: The Last 60 Years," *Parassitologia* 42 (2000): 69–85.

11. The impact of these investments are evident in Kerala's key social indicators. Compared to India as a whole, Kerala has a much lower infant mortality rate (13 vs.

84/1,000), crude birth rate (17 vs. 29/1,000), and crude death rate (6 vs. 9.2/1,000). It also has much higher literacy levels for both males (94 vs. 64 percent) and females (86 vs. 39 percent).

12. The repeated application of pesticides also led to behavioral adaptations on the part of certain species of anopheline mosquitoes. Malaria control programs sprayed pesticides on the walls of huts to kill resting mosquitoes. In some cases the irritant nature of the pesticide led vectors to rest out of doors. In other cases, the pesticide selected in favor of insects that chose to rest out of doors. Either way, the change in the mosquitoes' behavior reduced the effectiveness of hut spraying control strategies.

13. G. Chapin and R. Wasserstrom, "Agricultural Production and Malaria Resurgence in Central America and India," *Nature* 293, 5829 (1981): 181–85.

14. D. R. Roberts, L. L. Laughlin, et al., "DDT, Global Strategies, and a Malaria Control Crisis in South America," *Emerging Infectious Diseases* 3, 3 (1997): 295–302.

15. W. H. Wernsdorfer, "Epidemiology of Drug Resistance in Malaria," *Acta Tropica* 56, 2–3 (1994): 144.

16. The exceptions are the Americas north of the Panama Canal and some areas of southwestern Asia. Wernsdorfer postulates that these exceptions reflect the refractoriness of local malaria vectors to chloroquine-resistant strains of malaria. Ibid., 143–56.

17. D. Payne, "Did Medicated Salt Hasten the Spread of Chloroquine Resistance in Plasmodium falciparum?" *Parasitology Today* 4, 4 (1988): 112–15.

18. J. Verdrager, "Localized Permanent Epidemics: The Genesis of Chloroquine Resistance in *Plasmodium falciparum*," *Southeast Asian Journal of Tropical Medicine and Public Health* 26, 1 (1995): 23–28.

19. Ronald Ross, *The Prevention of Malaria* (London: John Murray, 1910).

20. G. M. Jeffrey, "Malaria Control in the Twentieth Century," *American Journal of Tropical Medicine and Hygiene* 23, 3 (1976): 361–71.

21. Javed Siddiqi, *World Health and World Politics* (Columbia: University of South Carolina Press, 1995), 141–46.

22. J. A. Najera, "Malaria Control: Achievements Problems and Strategies," Geneva, WHO, 1999 (WHO/MAL 99.1087), 67.

23. USAID 1973, Office of the Auditor General, Washington, D.C., Office of Audit, Report on Audit of the Malaria Eradication Programme, Audit No. 74-003, August 31, 1973.

24. WHO Archives, Geneva, Candau to Payne, August 6, 1962, M2/180/5.

25. Burton A. Weisbrod, *Economics of Public Health* (Philadelphia: University of Pennsylvania, 1961); Selma Mushkin, "Health as an Investment," *Journal of Political Economy,* suppl. (October 1962); Ansely J. Coale and Edgar M. Hoover, *Population Growth and Economic Development in Low-Income Countries: A Case Study of India's Prospects* (Princeton: Princeton University Press, 1958), 23; Robin Barlow, "The Economic Effects of Malaria Eradication," *Bureau of Public Health Economics, Research Notes,* no. 15 (Ann Arbor, 1968), 130.

26. Personal communication from Dr. Brandon Alan Kohrt, February 14, 1996.

27. C. Gramiccia and P. F. Beales, "The Recent History of Malaria Control and Eradication," in W. H. Wernsdorfer and I. McGregor, eds., *Malaria: Principles and Practice of Malariology* (Edinburgh: Churchill Livingstone, 1988).

28. J. Mouchet, S. Laventure, S. Blanchy, R. Fioramonti, A. Rakotonjanabelo, P. Rabarison, J. Sircoulon, and J. Roux, "The Reconquest of Madagascar Highlands by Malaria," *Bulletin de la Société de Pathologie Exotique* 90, 3 (1997): 162–68.

29. Southeast Asia Regional Organization (SEARO), WHO, "Malaria Situation in SEARO Countries," Geneva, 2003.

30. WHO, *Weekly Epidemiological Record* 74, 32 (1999): 265–72.

31. WHO, *World Health Report* (Geneva, 1999), 50.

Chapter 7. Malaria Realities

1. CDC, "Transmission of *Plasmodium vivax* Malaria—San Diego County, 1988 and 1989," *Morbidity and Mortality Review Weekly* 39, 6 (February 16, 1990): 91–94; Norma Mohr, *Malaria: Evolution of a Killer* (Seattle: Serif and Pixel Press, 2001); www.cdc.gov/malaria/control_prevention/mexico.htm.

2. See R. M. Packard, "Maize, Cattle, and Mosquitoes: The Political Economy of Malaria Epidemics in Colonial Swaziland," *Journal of African History* 25, 2 (1984): 189–212.

3. The following discussion is drawn from my earlier article: R. M. Packard, "Agricultural Development, Migrant Labor and the Resurgence of Malaria in Swaziland," *Social Science and Medicine* 22, 8 (1986): 861–67.

4. These conditions were cited to the commission investigating a strike at the Big Bend Sugar estates in 1963.

5. Donald R. Roberts, "DDT, Global Strategies and a Malaria Control Crisis in South America," *Emerging Infectious Diseases* 3, 3 (1997): 296.

6. For a discussion of Amazon colonization and its impact on malaria, see A. C. Marques, "Human Migration and the Spread of Malaria in Brazil," *Parasitology Today* 3, 6 (1987): 166–70; B. H. Singer and M. C. de Castro, "Agricultural Colonization and Malaria on the Amazon Frontier," *Annals of the New York Academy of Sciences,* 954 (2001): 184–222; Donald Sawyer, "Economic and Social Consequences of Malaria in New Colonization Projects in Brazil," *Social Science and Medicine* 37, 9 (1993): 1131–36.

7. Below a critical density, however, transmission could not be sustained. Robert Fontaine, McWilson Warren, and Jesse Hobbes, "Integrated Control Demonstration Project," unpublished project report, United States Public Health Service, n.d.

8. I had the pleasure of teaching the "malaria in El Salvador" story as a case study in a course on "Strategies in International Health" while I was at Emory University. I was part of a team that included several CDC researchers and program officers who had been involved in designing and implementing El Salvador's Integrated Malaria Program in the 1980s. The other members of the team were: Jesse Hobbes, Carlos Campbell, Frank Richards, Monica Parisse, and Trent Ruebush. The case was presented as an example on how to organize an integrated health program and stressed the importance of program flexibility. It was only after a student in the course asked

whether we had any evidence that the program was responsible for the decline in malaria in the 1980s that we began to think about alternative explanations.

9. Almost all of the 40 percent of the population, which retained access to land, owned small plots of less than 10 hectares. By 1971, 92 percent of all farms measured less than 10 hectares.

10. Personal communication from Dr. Carlos Campbell. Dr. Campbell participated in the Integrated Malaria Program as a U.S. Public Health officer in the 1980s.

11. Meredeth Turshen, *Privatizing Health Services in Africa* (New Brunswick: Rutgers University Press, 1999), 9.

12. Other examples are not hard to find. As we saw in chapter 3, Union troops fighting in the American Civil War suffered multiple episodes of the disease resulting in an incidence rate of 2,698 cases per 1,000 troops between May 1861 and June 1866. During this same period, the disease accounted for 10 percent of all deaths among Union troops. In World War I, the battle line between the Italian and Austro-Hungarian armies stretched for two and a half years across the Veneto plains, a notoriously malarious region of northeast Italy. Many of the troops on both sides came from regions that were malaria free. Fighting in mosquito-infested trenches and craters, these nonimmune troops suffered heavily from malaria. Conditions in the trenches were so bad that the Italian Third Army was known as the "malarial army." See Frank Snowden, *The Conquest of Malaria: Italy, 1900–1962* (New Haven: Yale University Press, 2006). American troops fighting in New Guinea and Guadalcanal in 1942–43 suffered similar levels of disease, though fewer deaths.

13. Colonial economic policies (described in chapter 4), the disadvantaged position of former colonies in global commodity markets, and ill-advised economic decisions by both newly independent governments and external development agencies have all contributed to the weak economies of many of the states in which military conflicts have occurred.

14. R. M. Garfield, E. Prado, et al., "Malaria in Nicaragua: Community-Based Control Efforts and the Impact of War," *International Journal of Epidemiology* 18, 2 (1989): 434–39.

15. S. Pitt et al., "War in Tajikistan and the Re-emergence of *Plasmodium falciparum,*" *Lancet* 352 (1998): 1279.

16. F. Checchi, J. Cox, et al., "Malaria Epidemics and Interventions: Kenya, Burundi, Southern Sudan, and Ethiopia, 1999–2004," *Emerging Infectious Diseases* 12, 10 (2006): 1477–85.

17. L. Roberts, C. Hale, F. Belyakdoumi, L. Cobey, R. Ondeko, M. Despines, et al., *Mortality in Eastern Democratic Republic of Congo* (New York: International Rescue Committee, 2001); B. Coghlan, R. J. Brennan, P. Ngoy, et al., "Mortality in the Democratic Republic of Congo: A Nationwide Survey," *Lancet* 367 (2006): 44–51.

18. M. Rowland and F. Nosten, "Malaria Epidemiology and Control in Refugee Camps and Complex Emergencies," *Annals of Tropical Medicine and Parasitology* 95, 8 (December 2001): 741- 54.

19. Médecins Sans Frontière Reports, "Burundi: Fever, Hunger and War," December 13, 2001, www.msf.org/msfinternational.

20. Goma Epidemiology Group, 1995; "Public Health Impact of Rwandan Refugee Crisis: What Happened in Goma, Zaire, in July, 1994?" *Lancet* 345 (1995): 339–404.

21. Peter A. Boland and Holly A. Williams, *Malaria Control during Mass Population Movements and Natural Disasters* (Washington, D.C.: National Academies Press, 2002), 11–13.

22. Neill Wright (UN High Commission for Refugees Regional Commissioner), "Population Displacement Health Challenges in Today's World," speech to the 8th National Rural Health Conference, Alice Springs, Northern Territory, Australia, March 10, 2005.

23. U.S. Committee for Refugees and Immigrants, *Statistical Issue* 25, 9 (December 31, 2004).

24. Paul Collier, *Breaking the Conflict Trap: Civil War and Development Policy*, 1 A, World Bank Policy Research Report (Oxford: World Bank and Oxford University Press, 2003), 39.

25. The presence of malaria among Afghan refugees entering into Pakistan has been hotly debated. Malaria control officials in Pakistan initially blamed increases in malaria within the NWFP to the introduction of malaria by Afghan refugees. This hypothesis was rejected by Mohammed Suleman of the University of Peshawar in Pakistan in "Malaria in Afghan Refugees in Pakistan," *Transactions of the Royal Society for Tropical Medicine and Hygiene* 81, 1 (1988): 44–47. Suleman's case rested on the age-specific distribution of malaria parasites among Afghan refugees and local populations. He concluded that malaria infections in adult refugees as well as their young children indicated a general lack of immunity to the disease. This, he argued, showed that the refugees had limited experience with malaria. In local Pakistan populations, by contrast, young children were infected but not adults, indicating the presence of acquired immunities and the existence of recurrent transmission within Pakistan. This difference in exposure was taken as proof that Afghan refugees did not introduce malaria into the country, but acquired their infections once they arrived. Yet, the age distribution of malaria among Afghan refugees is also consonant with the historical scenario described here, in which Afghani refugees had been only recently exposed to malaria after living for years under successful malaria control. Both sides in this debate have found support in subsequent publications. J. H. Kazmi and P. Kavita, "Disease and Dislocation: The Impact of Refugee Movements on the Geography of Malaria in NWFP, Pakistan," *Social Science & Medicine* 52, 7 (2001): 1043–55; M. Rowland, M. A. Rab, et al., "Afghan Refugees and the Temporal and Spatial Distribution of Malaria in Pakistan," *Social Science and Medicine* 55, 11 (2002): 2061–72. The disagreement appears to be fueled on one side by the need to defend the malaria control program from accusations of failure, and on the other by the wish to defend refugees from discrimination.

26. A study conducted among Afghan refugees found that proximity to cattle and to goats increases the subjects' chances of being bitten by anopheline mosquitoes. Man-biting by *Anopheles stephensi* rose by 38 percent (8–68 percent Confidence Interval) in the presence of a cow, and by 50 percent (16–84 percent CI) in the presence of two goats. S. Hewitt, M. Kamal, et al., "An Entomological Investigation of

the Likely Impact of Cattle Ownership on Malaria in an Afghan Refugee Camp in the North West Frontier Province of Pakistan," *Medical and Veterinary Entomology* 8, 2 (1994): 160–64. The spraying of cattle in six refugee camps in 1995 and 1996, significantly reduced malaria transmission within the camps. M. D. Rowland et al., "Control of Malaria in Pakistan by Applying Deltamethrin Insecticide to Cattle: A Community-Randomised Trial," *Lancet* (British edition) 357, 9271 (2001): 1837–41.

27. The disparity between NWFP rates and those for the rest of the country would also seem to undermine the argument that some have made, that the rise in malaria, and particularly falciparum cases during the 1980s, was due to the spread of chloroquine-resistant strains of *falciparum*. Susceptibility surveys conducted among sentinel populations throughout Pakistan indicated that chloroquine resistance first appeared in 1981 and subsequently spread rapidly throughout the malarious parts of the country. There is no evidence that it was more prevalent in the NWFP than elsewhere in the country. Similarly, while climate change—wetter rainy seasons and increased temperatures—may have contributed to the rise in malaria incidence in all of Pakistan during the 1980s, it would not account for the greater rise in NWFP.

28. W. K. Reisen and P. F. L. Boreham, "Estimates of Malaria Vectorial Capacity for *Anopheles culicifacies* and *Anopheles stephensi* in Rural Punjab Province, Pakistan," *Journal of Medical Entomology* 19, 1 (1982): 98–103.

29. F. O. ter Kuile, M. E. Parise, et al., "The Burden of Co-infection with Human Immunodeficiency Virus Type 1 and Malaria in Pregnant Women in Sub-Saharan Africa," *American Journal of Tropical Medicine and Hygiene* 71, 2 (suppl.) (2004): 41–54.

30. These figures are rough estimates given the poor quality of malaria case reporting in much of Africa. E. L. Korenromp, "Malaria Attributable to the HIV-1 Epidemic, Sub-Saharan Africa," *Emerging Infectious Diseases* 11, 9 (2005): 1410–19.

31. John Gallup and Jeffrey Sachs, "The Economic Burden of Malaria," *American Journal of Tropical Medicine and Hygiene* 64, 1–2 (2001): 85–96.

32. WHO/UNICEF, *Africa Malaria Report* (Geneva, 2003).

33. Vinay Kamat, "Negotiating Illness and Misfortune in Post-Socialist Dar-es-Salaam" (Ph.D. dissertation, Emory University, 2004). The study revealed that clinics provided chloroquine for free. But if this did not work, patients had to pay for second-line drugs. Because chloroquine resistance was widespread, mothers recognized that taking their child to clinic would entail a cost.

34. Anchalee Singhanetra-Renard, "Population Movement, Socio-Economic Behavior and the Transmission of Malaria in Northern Thailand," *Southeast Asian Journal of Tropical Medicine and Public Health* 17, 3 (1986): 396–405.

35. Integrated Regional Information Networks, Africa, UN Office for the Coordination of Humanitarian Affairs, May 14, 2003. In an even more tragic occurrence, the Zimbabwean government forcefully removed hundreds of thousands of poor Zimbabweans, mostly unemployed people, from Harare and other major urban areas in the name of an urban renewal and slum clearance program known as Operation Murambatsvina—which is Shona for "get rid of trash." The removals forced many people to relocate in the rural areas of the country. The United Nations estimated that up to 360,000 people (more than 2 percent of Zimbabwe's population)

were evicted from their homes. Reports from the countryside indicated that those removed from the cities were highly vulnerable to malaria and, like the illegal gold panners, possessed the resources neither to protect themselves from infection nor to obtain treatment once they became sick. One cross-border trader from Gokwe, a malarial area, told a local newspaper, the *Zimbabwean* (September 2, 2005): "We have witnessed many deaths from people who were evicted from urban areas. Malaria is even worse than HIV/AIDS. At least that disease gives you time to seek medication. But malaria kills very quickly." The exact motivation for the removals has been hotly debated, with government opponents seeing it as a means of removing groups who supported the opposition Movement for Democratic Change, in the 2005 elections. But, what is important was that the people removed by the operation were a product of the country's devastated economy. It was their poverty that led them to oppose the ZANU-PF government in the 2005 election; whether it was this opposition, or the government's desire to clean up the country's urban areas, their poverty ultimately created the conditions of exposure that caused them to succumb to malaria. Michael Wines, "Zimbabwe's 'Clean-Up' Takes a Vast Human Toll," *New York Times,* June 11, 2005.

36. Gladys Tettah, "Rapid Assessment of Antimalarial Drug Availability and Use in Nigeria, February to March 2004," Rational Pharmaceutical Plus Program, USAID, Arlington, Virginia, April 2005.

37. WHO/UNICEF, *Africa Malaria Report,* 2003.

38. A. A. Amin and R. W. Snow, "Brands, Costs and Registration Status of Antimalarial Drugs in the Kenyan Retail Sector," *Malaria Journal* 4, 1 (2005): 36.

39. Public Broadcasting System, "Malaria: Fever Wars," April 6, 2006.

40. World Bank, *World Development Indicators* (Washington, D.C., 2006).

41. By 1970 the infant mortality rate in Zambia was 108 per 1,000. Life expectancy was around 50 years, and 42 percent of people 15 years and older were literate. The ratio of doctors to population in 1965 was 1:11,300.

42. World Development Movement, *Zambia: Condemned to Debt* (April 2004), 9.

43. Oxfam, *Cultivating Poverty: The Impact of US Cotton Subsidies on Africa,* Oxfam Briefing Paper 30 (New York: Oxfam International, 2002), 17–21. For countries like Benin, Burkina Faso, and Mali, which produce much more cotton and depend on it as a major source of export earning, the losses have been much greater. Oxfam estimated the following losses: Burkino Faso, 1 percent of GDP and 12 percent of export earnings; Mali, 1.7 percent of GDP and 8 percent of export earnings; and Benin, 1.4 percent of GDP and 9 percent of export earnings.

44. Jorge F. Balat and Guido G. Porto, "The WTO Doha Round Cotton Sector Dynamics and Poverty Trends in Zambia," in Thomas W. Hertel and L. Alan Winters, eds., *Putting Development Back into the Doha Agreement, Poverty Impacts of a WTO Agreement* (Washington, D.C.: World Bank, 2005).

45. Much more substantial debt relief came in June 2005, when International Monetary Fund, World Bank, and Western donors cut Zambia's foreign debt to $502 million from an estimated $7.2 billion.

46. "Zambia's Nursing Brain Drain," *BBC Saturday,* September 23, 2006, http://news.bbc.co.uk/1/hi/world/africa/5362958.stm.

47. World Bank, *Health Sector Support Project: Zambia,* Staff Appraisal Report, World Bank (Washington, D.C., October 14, 1994). Whether patients stopped attending clinics because of the cost, or because the clinics did not provide quality service and patients refused to pay for poor service is unclear.

48. Yann Derriennic et al., "Impact of User Fee Waiver Pilot on Health Seeking Behavior of Vulnerable Populations in Kafue District, Zambia," paper presented at American Public Health Association, Philadelphia, December 2005.

49. Only 15,000 people were being treated with antiretroviral therapy (ART). A rapid assessment of the Zambian ART program identified several important constraints including: inadequate human resources for testing, counseling, and treatment-related care; gaps in supply of drugs in the public sector; lack of adequate logistic or supply chain systems; and the high cost of ART to patients. Many of the weaknesses of the ART program reflected the overall weaknesses of the health system. As a British doctor, who visited Zambia in 2000, noted, "It is not simply anti-retroviral agents which are inaccessible to many but treatment for infections and symptoms, health education, distribution of condoms and effective health promotion." World Development Movement, *Zambia: Condemned to Debt,* 40.

50. Anonymous, *Malaria in Zambia: Situation Analysis, May 2000,* National Malaria Control Centre, Central Board of Health (Lusaka, Zambia, 2000).

51. Ibid., 24.

52. More recently, grants from the Global Fund have dramatically increased access to Coartem. Personal communication from Dr. Kent Campbell, February 26, 2006.

53. See James Ferguson, *Expectations of Modernity: Myths and Meaning of Urban Life on the Zambian Copperbelt* (Berkeley: University of California Press, 1999), for a description of this process.

54. Anonymous, *Malaria in Zambia,* 6–7.

55. Central Statistical Office, *Zambia Demographic and Health Survey, 2001–2002* (Lusaka, Zambia, 2003).

Chapter 8. Rolling Back Malaria

1. The strategy also called for "the creation of national and local capacities for assessing malaria situations and selecting appropriate measures that are aimed at reducing or preventing the disease problem in the community rather than being concentrated on reducing parasite rates in the populations, as was too often the case in the past."

2. World Health Organization, *Global Strategy for Malaria Control* (Geneva: WHO, 1993), viii.

3. Kenneth J. Arrow, Claire Panosian, and Hellen Gelband, eds., *Saving Lives, Buying Time: Economics of Malaria Drugs in an Age of Resistance* (Washington, D.C.: National Academies Press, 2004), 202–4.

4. www.cdc.gov/malaria/pregnancy.htm.

5. The revised RBM Global Strategic Plan for 2005–2015 calls in addition for 80 percent coverage of at-risk populations with appropriate antimalarial protections, and for 80 percent of malaria patients to be diagnosed and treated with effective antimalarial treatment within one day of the onset of illness.

6. L. Molineaux and G. Gramiccia, *The Garki Project* (Geneva: WHO, 1980).

7. Norvartis's partnership with WHO provided the company with a dominant market position in the worldwide sales of ACTs. More recently, a partnership has been created between the Global Funds for AIDS, Tuberculosis and Malaria, and major private corporations, including American Express, Reebok, and the GAP. The private companies will produce a line of products labeled "RED." One percent of the sales from these products will go to the Global Fund.

8. Personal communication from Dr. Clive Shiff, entomologist, Bloomberg School of Public Health, May 2006.

9. Kenneth J. Arrow, Hellen Gelband, and Dean T. Jamison, "Making Anti-malarial Agents Available in Africa," *New England Journal of Medicine* 353 (2005): 333–35.

10. Interview with Dr. Protik Basu, Public Health Specialist, Malaria Team, World Bank, Baltimore, January 4, 2005.

11. Cathy Green, *External Evaluation of Roll Back Malaria Program: Report of RBM Stakeholder Interviews* (Quebec: Health Partners International, March 2002), 25. It is unclear to what extent African countries understood what the World Bank meant when it pledged hundreds of millions of dollars to combat malaria in Africa at the Abuja summit. From the bank's perspective, the failure of funding came from the unwillingness of countries to devote International Development Association credits to malaria.

12. www.afronets.org/archive/199910/msg00086.php. The UNDP was charged with creating capacity for integration of malaria-related action into national poverty eradication policies, strategies and programs; strengthening, through sustainable human development activities, the balance of action among the state, private sector, civil society, and the communities themselves, to ensure that people have access to basic social services and productive assets; and working through the UN Resident Coordinator system to encourage collaborative programming in support of intersectoral action and resource mobilization.

13. See, for example, P.-O. Pehrson et al., "Is the Working Capacity of Liberian Industrial Workers Increased by Regular Malaria Prophylaxis," *Annals of Tropical Medicine and Parasitology* 78 (1984): 453–58; J. J. Brohult et al., "The Working Capacity of Liberian Males: A Comparison of Urban and Rural Populations in Relationship to Malaria," *American Journal of Tropical Medicine and Parasitology* 75 (1981): 487–94; M. Audibert, "Agricultural Non-wage Production and Health Status: A Case Study in a Tropical Milieu," *Journal of Development Economics* 24 (1986): 275–91; El Tahir Mohamed Nur, "The Impact of Malaria on Labour Use and Efficiency in the Sudan," *Social Science & Medicine* 37, 9 (November 1993): 1115–19.

14. John Luke Gallup and Jeffrey Sachs, "The Economic Burden of Malaria," *American Journal of Tropical Medicine and Hygiene* 64, 1, 2 (2001) 85–96.

15. Jeffrey Sachs and Pia Malaney, "The Economic and Social Burden of Malaria," *Nature* 415, 7 (February 2002): 680–85.

16. Sachs's calculations received little in the way of serious critical appraisal by economists. His two articles were not cited in any economic or development journals prior to 2006. They were, however, cited by nearly everyone else doing work on

malaria as justification for investments in malaria research and control. His calculations were also used to evaluate the relative economic impact of malaria control measures compared to other types of investments at the Copenhagen Consensus meeting in 2004. The paper arguing for the economic impact of malaria control was part of a larger evaluation of the economic impact of communicable diseases, authored by Anne Mills and Sam Shillcut. The authors acknowledged that there were limitations to the types of analysis employed by Gallup and Sachs, commenting that there was a need for a deeper understanding of the mechanisms by which malaria-affected households and economies and concluded that, without this deeper understanding, it was difficult to evaluate existing estimates, both micro and macro. Nonetheless, they employed Gallup and Sachs's estimates along with those of a similar analysis by McCarthy, Wolf, and Wu, to calculate the economic benefits of successfully achieving the RBM goals. On the basis of this analysis, the Copenhagen Consensus ranked the implementation of recent approaches to malaria control as the fourth most cost-effective intervention for generating economic growth in sub-Saharan Africa. Bjorn Lomborg, ed., *Global Crises, Global Solutions* (Cambridge: Cambridge University Press, 2004), 605.

17. If they were attempting to curtail government expenditures under a World Bank–directed Poverty Reduction program, however, they would be under pressure to require individuals to purchase nets.

18. In September 2006, Dr. Arata Kochi, head of WHO's Global Malaria Program announced a change in WHO policy on indoor residual spraying. The Global Fund for AIDS, Tuberculosis, and Malaria has funded spraying programs in several countries.

19. Dominic Kwiatkowski and Kevin Marsh, "Development of a Malaria Vaccine," *Lancet* 350 (December 6, 1997): 1699.

20. This was based on a range of surveys conducted between 1999 and 2004.

21. Erin Eckert, Ani E. Hyslop, and Robert Snow, "Estimating the Feasibility of Implementing Intermittent Preventive Treatment (IPT) for Malaria through Routine Antenatal Care in Africa," paper presented at the American Public Health Association annual meeting, Philadelphia, December 2005.

22. *Tanzania: Demographic and Health Survey, 2004–2005* (Tanzania: National Bureau of Statistics; ORC Macro, 2005).

23. H. L. Guyatt, S. A. Ochola, and R. W. Snow, "Too Poor to Pay: Charging for Insecticide-Treated Bednets in Highland Kenya" *Tropical Medicine and International Health* 7, 10 (2002): 846–80; O. Onwujekwe et al., "Hypothetical and Actual Willingness to Pay for Insecticide-Treated Nets in Five Nigerian Communities," *Tropical Medicine and International Health* 6, 7 (2001): 545–53.

24. PSI Kenya Brief, www.rollbackmalaria.org/docs/Kenya_brief.pdf.

25. WHO, *World Malaria Report* (Geneva, 2005), 24.

26. Ibid., 26.

27. S. Patrick Kachur et al., "Maintenance and Sustained Use of Insecticide Treated Bed Nets Three Years after a Controlled Trial in Western Kenya," *Tropical Medicine and International Health* 4, 11 (1999): 728–35.

28. P. J. Winch, A. M. Makemba, et al., "Seasonal Variation in the Perceived

Risk of Malaria: Implications for the Promotion of Insecticide-Impregnated Bed Nets," *Social Science and Medicine* 39, 1 (1994): 63–75.

29. Numerous studies support this conclusion: Dyna Arhin-Tenkorang, "Mobilizing Resources for Health: The Case for User Fees Revisited," Harvard Center for International Development Working Paper No. 81 (December 2001); B. Jacobs and N. Price, "The Impact of the Introduction of User Fees at a District Hospital in Cambodia," *Health Policy and Planning* 19, 5 (2004): 310–21; M. G. Segall et al., "Economic Transition Should Come with a Health Warning: The Case of Vietnam," *Journal of Epidemiology and Community Health* 56, 7 (2002): 497- 505. By contrast, J. H. Bratt, M. A. Weaver, et al., "The Impact of Price Changes on Demand for Family Planning and Reproductive Health Services in Ecuador," *Health Policy and Planning* 17, 3 (2002): 281–87, reported that price increases had little effect on the use of reproductive health services by the poor.

30. CDC, "Malaria Control in Uganda—Towards the Abuja Targets," www.cdc.gov/malaria/control_prevention/uganda.htm.

31. WHO, "Malawi Roll Back Malaria Consultative Mission, Final Report," August 23, 2004, 13.

32. World Bank, *Global Strategy and Booster Program* (Washington, D.C.: World Bank, 2005), 7.

33. World Bank, *Zambia Malaria Booster Project: Project Information Document* (Washington, D.C., November 2004), 2. The actual percentage change that the bank hoped to achieve was unclear. Three different bank documents related to the Zambia Booster Project, released in the fall of 2005, gave three different sets of percentage changes.

34. Interview with Dr. Protik Basu, Baltimore, March 16, 2006.

35. Government of Zambia, *Demographic and Health Survey, 2001–2002* (Lusaka, 2003). The survey found that of women who have given birth over the five years preceding the survey, 9 out of 10 had received antenatal care from a health professional.

36. Takatoshi Kato, "Post-Debt Relief Opportunities and Challenges for Zambia," *Post* (Zambia), August 3, 2005.

37. The World Bank is committed to what is called sector-wide approaches, or SWAps, to health system development. Depending on to whom one speaks, SWAps involve one of two strategies. The first involves the placement of the financial contributions of various donors into a single "basket." These funds are then available to the country involved to spend on activities identified in its own national health plan. The second interpretation involves donors investing in their specific health activities in a coordinated fashion, so that the various programs are integrated into the country's health plans. SWAps in Zambia and other African countries have been largely about process. There is little evidence that these have contributed to significant improvements in the overall health systems. The bank's review of the first year of the booster program included a section on "Building on the Health System Constraints." It identified the constraints imposed on malaria control programs by the weaknesses of the local health system, but said little about what had been achieved to correct these weaknesses. The section ended with the sentence: "The Bank is work-

ing with partners to address these issues in the context of the overall health dialogue at country level." World Bank, "Booster Program for Malaria Control in Africa: One Year Later: Progress and Challenges," document prepared for Regional Booster Program Meeting, Dakar, September 12–15, 2006.

38. C. C. Loila, C. J. da Silva, and P. L. Tauil, "Malaria Control in Brazil, 1965–2001" (in Portuguese), *Revista Panamericana de Salud Pública* 4 (2002): 235–44.

39. World Bank, "Malaria and Reproductive Health Project for Eritrea," 66, Project Appraisal Document, proposed grant for HIV/AIDS/STI, TB, May 31, 2005.

40. Interview with Dr. Protik Basu, March 16, 2006.

41. World Bank, "Malaria and Reproductive Health Project for Eritrea," 25.

42. Lieutenant Colonel S. P. James, following a four-month tour of Kenya and Uganda in 1929, concluded that acquired immunity provided African adults with protection from malaria and that control measures should be limited to a therapeutic program aimed at attenuating clinical symptoms of disease rather than radically eliminating the parasite, because the latter would undermine the natural immunity acquired by recurrent infections. He stated at a conference of the Malaria Committee of the League of Nations in 1933 that, "it would be extremely unwise, in certain malarious countries (Dark Africa for example) to intervene radically in the natural process, that to which the natives are immunized against the disease." D. Bagster Wilson, who spent 30 years researching malaria in East Africa beginning in the late 1920s, took up James's arguments. Like James, he promoted the idea that malaria control measures, such as spraying and drainage, should be limited to populations of nonimmune Africans. However, he stated that it was critically important to identify various levels of malaria transmission and distinguish immune from nonimmune populations. Bagster Wilson, along with P. C. C. Garnham and N. L. Swellengrebel, repeated these arguments in opposition to the extension of malaria control programs, using DDT, to highly endemic areas of Equatorial Africa in the 1950s. They argued at a malaria meeting held in Kampala in 1950 that unless malaria control measures could be sustained, they should not begin, for eliminating or greatly reducing transmission would undermine the acquired immunity of the populations. M. Malowany, "Unfinished Agendas: Writing the History of Medicine in Africa," *African Affairs* 99 (2000): 342; M. J. Dobson, M. Malowany, and R. W. Snow, "Malaria Control in East Africa: The Kampala Conference and the Pare-Taveta Scheme: A Meeting of Common and High Ground," *Parassitologia* 42 (2000): 149–66. Paul Russell of the Rockefeller Foundation, who was a strong advocate of vector control, and George Macdonald, who developed the mathematical model upon which the strategy of malaria eradication was based, argued that malaria control measures should be applied wherever the disease occurred, pointing to the reduction in childhood mortality that could be achieved in populations in which adults possessed acquired immunity.

43. N. Askjaer, C. Maxwell, et al., "Insecticide-Treated Bed Nets Reduce Plasma Antibody Levels and Limit the Repertoire of Antibodies to *Plasmodium falciparum* Variant Surface Antigens," *Clinical and Diagnostic Laboratory Immunology* 8, 1 (2001): 1289–91. See also S. K. Kariuki, A. A. Lal, et al., "Effects of Permethrin-Treated Bed

Nets on Immunity to Malaria in Western Kenya II: Antibody Responses in Young Children in an Area of Intense Malaria Transmission," *American Journal of Tropical Medicine and Hygiene* 68, 4 (suppl.) (2003): 108–14; W. G. Metzger, C. A. Maxwell, et al., "Anti-sporozoite Immunity and Impregnated Bednets in Tanzanian Villages," *Annals of Tropical Medicine and Parasitology* 9, 6 (1998): 727–29.

44. K. A. Lindblade, T. P. Eisele, et al., "Sustainability of Reductions in Malaria Transmission and Infant Mortality in Western Kenya with Use of Insecticide-Treated Bednets: 4 to 6 Years of Follow-Up," *Journal of the American Medical Association* 291, 21 (2004): 2571–80; C. A. Maxwell, E. Msuya, et al., "Effect of Community-wide Use of Insecticide-Treated Nets for 3–4 Years on Malarial Morbidity in Tanzania," *Tropical Medicine and International Health* 7, 12 (2002): 1003–8; T. P. Eisele, K. A. Lindblade, et al., "Effect of Sustained Insecticide-Treated Bed Net Use on All-Cause Child Mortality in an Area of Intense Perennial Malaria Transmission in Western Kenya," *American Journal of Tropical Medicine and Hygiene* 73, 1 (2005): 149–56.

45. Interview with Carlos Kent Campbell, January 3, 2006.

46. F. Chandre, F. Darrier, et al., "Status of Pyrethroid Resistance in *Anopheles gambiae* sensu lato," *Bulletin of the World Health Organization* 77, 3 (1999): 230–34; A. E. Yawson, P. J. McCall, et al., "Species Abundance and Insecticide Resistance of *Anopheles gambiae* in Selected Areas of Ghana and Burkina Faso," *Medical and Veterinary Entomology* 18, 4 (2004): 372–77.

47. A. N. Asidi, R. N'Guessan, et al., "Experimental Hut Evaluation of Bednets Treated with an Organophosphate (chlorpyrifos-methyl) or a Pyrethroid (lambdacyhalothrin) Alone and in Combination against Insecticide-Resistant *Anopheles gambiae* and *Culex quinquefasciatus* Mosquitoes," *Malaria Journal* 4, 1 (2005): 25; J. Etang, F. Chandre, et al., "Reduced Bio-efficacy of Permethrin EC Impregnated Bednets against an *Anopheles gambiae* Strain with Oxidase-Based Pyrethroid Tolerance," *Malaria Journal* 3 (2004): 46.

48. V. Corbel, J. M. Hougard, et al., "Evidence for Selection of Insecticide Resistance Due to Insensitive Acetylcholinesterase by Carbamate-Treated Nets in *Anopheles gambiae* s.s. (Diptera: Culicidae) from Cote d'Ivoire," *Journal of Medical Entomology* 40, 6 (2003): 985–88.

49. The threat of resistance arising from this practice led the director of WHO's malaria control program, Dr. Arata Kochi, to pressure these companies to cease their practice.

50. By the end of 2006, Zambia's external debt had been reduced to $500 million.

Conclusion. Ecology and Policy

1. World Health Organization, *Global Strategy for Malaria Control* (Geneva: WHO, 1993), viii.

INDEX

Afghanistan, refugees from, 196–99, 281n25, 281–82n26
Africa: HIV infection in, 200–201; malaria in, 13, 16, 19, 20, 22–24, 26–31, 102–9, 154, 175, 200–201, 219–20, 230–35, 288n42; migrant workers from, 89–90. *See also* Burundi; Congo; Egypt; Kenya; Malawi; Nigeria; South Africa; Swaziland; Uganda; Zambia; Zimbabwe
Agreste (Brazil), 92
Agricultural Adjustment Act (U.S.), 77–78
agricultural development schemes and malaria, 84–91, 109–10, 179–93; in Brazil, 184–87; in El Salvador, 187–93; in Punjab, 98–102; in South Africa, 109; in Swaziland, 179–87
agricultural transformations and malaria, 11–15, 28, 37, in American South, 68–78; in Illinois, 64–65; in the Roman Campagna, 42–44; in Sardinia, 40–41; in South Carolina, 56–61; in southeastern England, 48–50; in southern Italy, 79–81, 82
agriculture. *See* forest agriculture; plantation agriculture
AIDS: and malaria, 199–201; in Zambia, 212, 241
Amazon: colonization of, 185–86; malaria control in, 240–41; migration to, 95
Amazon Basin Malaria Control Project, 240–41
anemia, 73, 219
American South: black farmers in, 71–72; malaria in, 53–61, 70–75, 77
Annand, Percy, 275n41
Anopheles. See mosquitoes; *and names of species*
Anopheles albimanus, 25, 163, 188, 190, 191
Anopheles arabiensis, 260n13, 269n6
Anopheles atroparvus, 50, 52, 53, 54

Anopheles crucians, 57, 265n43, 265–66n44
Anopheles culicifacies, 163, 198
Anopheles darlingi, 185, 186
Anopheles dirus, 166
Anopheles fluviatilis, 163
Anopheles funestus, 108, 269n6
Anopheles gambiae, 25–28, 96, 105, 108, 154, 260n13, 275n35; breeding sites for, 26, 27, 93, 109, 215, 260n14; eradication of, 138–39; invasion of Brazil, 91–96; pyrethroid resistance in, 242–43; in Swaziland, 181, 182, 183
Anopheles hermsi, 177
Anopheles labranchiae, 34, 40, 143
Anopheles maculipennis, 7, 52, 125, 137
Anopheles messeae, 53
Anopheles philippinensis, 4, 7, 8
Anopheles pseudopunctipennis, 133
Anopheles quadrimaculatus, 6, 7, 55, 56, 57, 60, 61, 65, 73, 75, 125, 265n41; and the TVA, 132–33
Anopheles sacharovi, 34
Anopheles stephensi, 87, 198
anophelism without malaria, 125, 136–38
antimalarial drugs, 233; availability of, 203–4; during pregnancy, 219; resistance to, 164–67, 176, 202, 213, 243. *See also* artemisinin; atabrine; chloroquine; Coartem; primaquine; quinine; sulphadoxine pyrimethamine
antiretroviral therapy (ART), 284n49
Archangel, Russia, malaria epidemic in, 1–3, 7, 8, 9, 10
Argentina, 133–34
artemisinin, 114, 243
artemether-lumefantrine. *See* Coartem
Asia. *See* Central Asia; Cambodia; China; India; Pakistan; Southeast Asia; Thailand
atabrine, 140
Atomic Energy Commission, 149

Index